Integrated Science

Volume 28

The **Integrated Science** Series aims to publish the most relevant and novel research in all areas of Formal Sciences, Physical and Chemical Sciences, Biological Sciences, Medical Sciences, and Social Sciences. We are especially focused on the research involving the integration of two of more academic fields offering an innovative view, which is one of the main focuses of Universal Scientific Education and Research Network (USERN), science without borders.

Integrated Science is committed to upholding the integrity of the scientific record and will follow the Committee on Publication Ethics (COPE) guidelines on how to deal with potential acts of misconduct and correcting the literature.

Lílian Santos

Governance of Human Gene Editing and Transhumanism

Tasks for Today's Bioethics

Springer

Lílian Santos
UNESCO Chair in Bioethics and Human
Rights—Pontifical Athenaeum Regina
Apostolorum
Rome, Italy

Catholic Institute of Technology
Castel Gandolfo, Italy

ISSN 2662-9461 ISSN 2662-947X (electronic)
Integrated Science
ISBN 978-3-031-70862-6 ISBN 978-3-031-70863-3 (eBook)
https://doi.org/10.1007/978-3-031-70863-3

This Springer imprint is published by the registered company Springer Nature Switzerland AG
The registered company address is: Gewerbestrasse 11, 6330 Cham, Switzerland

Foreword

It is my pleasure to write this foreword for the book *Governance of Human Gene Editing and Transhumanism: Tasks for Today's Bioethics*. Lílian Santos, whose doctoral thesis I had the privilege of supervising, has meticulously crafted this work based on her extensive research.

Our journey was not without challenges. When Lílian first approached me, I was recovering from COVID-19 in the hospital. Fortunately, I survived the dreadful disease to supervise her dissertation. Over the years, we engaged in stimulating discussions, even when our opinions diverged. These debates encouraged us to delve deeper into the critical topics at hand.

Lílian faced initial hurdles due to the expansive scope of her subject matter, which encompassed gene editing, governance, and Transhumanism. Undeterred, her unwavering vision and dedication allowed her to develop a coherent narrative. Another obstacle was identifying the right method to objectively describe the relationship between the governance of gene editing and Transhumanism. Through perseverance, Lílian discovered and learned to use the method of Reflective Thematic Analysis—a significant breakthrough in her research.

This report is important for two reasons. First, the method described here is an invaluable tool, first used in a major bioethical work. Bioethicists frequently make substantial claims about relationships between topics without a clear process to demonstrate such relationships. The method described in this book will help students and researchers objectively investigate other topics in Bioethics.

Second, the subject matter described in this volume is significant. This book discusses the collision of ideas in the realms of Transhumanism and the governance of human gene editing. After providing evidence of common themes between these two areas, Lílian constructively engages the tasks for contemporary Bioethics regarding topics in which different interests tend to collide and accusations of ideological tendencies predominate.

The proactive approach of this work transcends prejudices. The author looks at the future of Bioethics on grounds that could be shared by conservatives and liberals, concentrating on scientific progress and its role in human flourishing while also warning about possible deviations.

I trust that readers will find both the subject matter and Lilian's approach refreshing, devoid of the usual polemics frequently present in Bioethics discussions, fostering meaningful dialogue in the field.

Rome, Italy
June 2024

Prof. Joseph Tham

Acknowledgements

I am extremely grateful for **God**'s guidance in completing this work, which is largely based on my doctoral dissertation in Bioethics, concluded in 2023. I thank my family, especially my parents **Jorge and Vânia**, who provided me with the initial education and always encouraged me to pursue this path.

I warmly thank my thesis adviser, **Prof. Joseph Tham**, who guided me, stimulated honesty and creativity in this research, and made my attendance possible at the Third International Summit on Human Genome Editing in London in 2023. I want to thank the second reader, **Prof. Michael Baggot**, who shared my interest in Transhumanism and supported my research with his detailed review. I am also grateful to **Prof. Ana María Yévenes** for opening up my horizons by introducing me to possible research methodologies used by the social sciences. I thank **Sydney Hyndman** for proofreading this book.

I am grateful to my Institutions, the **UNESCO Chair in Bioethics and Human Rights, the Pontifical Athenaeum Regina Apostolorum**, and the **Catholic Institute of Technology** for supporting my research. I thank my fellow faculty members for the teamwork and stimulating conversations. I also thank my students, who helped me to improve the content and present it with greater clarity.

I am deeply grateful to the Chairholder of the UNESCO Chair in Bioethics and Human Rights, **Prof. Alberto García**, who has long inspired me with his proactivity in this field. Also, I am grateful to the many partners and scholars I have met in my fellow faculty members time with the UNESCO Chair. These encounters have broadened my horizons, kept me updated on contemporary bioethical developments, and taught me the importance of dialogue and cooperation in a pluralistic society.

Introduction to the Global Governance of Human Genome Editing and Transhumanism

Few technologies touch our sense of identity as profoundly as gene editing. Thanks to tools such as CRISPR-Cas9 and Prime, it is now more precise, quicker, and cheaper to alter an organism's genome. Although the most advanced gene editing technologies continue to struggle with some technical issues, their existence represents a major achievement relative to past decades. There are reasons to be excited about the future possibilities that these technologies have to offer.

Which applications of gene editing are positive for humanity and the environment? Which are negative? Which applications continue to pose ethical challenges? How can we be prudent without letting fear block acceptable applications? How should we use gene editing to promote human health and improve our quality of life? How do we avoid the technical and social risks associated with such applications? Who are the decision-makers? Is global governance of genome editing feasible? How can we manage gene editing in a better way? Can the diverse stakeholders of our pluralistic society agree?

Interactions, decisions, and governance of technologies always occur in a historical context. The global governance of human genome editing (hereafter HGE) emerges at a time when some societies are engaging in transhumanist proposals. Transhumanist ideas promote the use of technology to profoundly improve human capabilities, thereby allowing individuals to shape themselves according to personal desire. This intellectual and cultural movement suggests a self-directed evolution. Transhumanist ideas may impact the global governance of new technologies. Transhumanist goals are not simply a futuristic dream of a couple of thinkers. These goals are translated into concrete work and projects which are conducted by very intelligent individuals. What is their purpose? To achieve and make *widely available* the so-called *safe and non-coercive enhancement through technology*.[1] Ideas such as enhancement and wish-fulfilment medicine are already more part of our mental framework than we probably realize.

[1] Cf. N. Bostrom, "A History of Transhumanist Thought", in M. Rectenwald—L. Carl (ed.), *Academic Writing Across the Disciplines*, Pearson Longman, New York 2011, 12.

This research project began with two questions: Are there commonalities between Transhumanism and the global governance of human genome editing? If yes, then what are the implications for Bioethics?

The first part of this book is dedicated to explore the common themes between Transhumanism and the plan for global governance of HGE. Part I synthesizes the excerpt from my doctoral dissertation in Bioethics titled *Common Themes Between Humanity+ Original Documents on Transhumanism and WHO Documents on the Global Governance of Human Genome Editing. Reflexive Thematic Analysis and Implications for Bioethics.*[2] Chapter 1 addresses the global governance of HGE, considering particularly the three documents published by the World Health Organization on this topic.[3] The texts were published in July 2021 and constitute the only framework for the global governance of HGE to date. Chapter 2 addresses the basics of Transhumanism and its original documents[4] according to Humanity+ (formerly known as the World Transhumanist Association). Chapter 3 reports on the application of the Reflexive Thematic Analysis (RTA) method to identify common themes between Transhumanism and the global governance of HGE in the above-mentioned documents.

The second part of this book is dedicated to the tasks for Bioethics, considering the common themes found in the RTA. Chapter 4 explains why current Bioethics must address the new possibilities, desires, and questions emerging with new technologies such as HGE. Chapter 5 makes explicit the limitations of the paradigm "therapy (permitted) versus enhancement (forbidden)". Chapter 6 presents work toward a more suitable ethical framework, moving from "therapy versus enhancement". This book contributes to the issue by offering the principle of proportionality in Chap. 7. Chapter 8 addresses the role of Bioethics in the global governance of HGE, as an ethics that goes beyond procedures and enlightens the key stakeholders. It also addresses some values and acknowledges

---

[2] Cf. L. SANTOS, *Common Themes Between Humanity+ Original Documents on Transhumanism and WHO Documents on the Global Governance of Human Genome Editing. Reflexive Thematic Analysis and Implications for Bioethics.*, Ateneo Pontificio Regina Apostolorum, Rome 2023.

[3] WHO EXPERT ADVISORY COMMITTEE ON DEVELOPING GLOBAL STANDARDS FOR GOVERNANCE AND OVERSIGHT OF HUMAN GENOME EDITING, *Human genome editing: a framework for governance*, WHO, Geneva 2021, in https://www.who.int/publications/i/item/9789240030060 [13-7-2021]; WHO EXPERT ADVISORY COMMITTEE ON DEVELOPING GLOBAL STANDARDS FOR GOVERNANCE AND OVERSIGHT OF HUMAN GENOME EDITING, *Human genome editing: position paper*, WHO, Geneva 2021, in https://www.who.int/publications-detail-redirect/9789240030404 [13-7-2021]; WHO EXPERT ADVISORY COMMITTEE ON DEVELOPING GLOBAL STANDARDS FOR GOVERNANCE AND OVERSIGHT OF HUMAN GENOME EDITING, *Human genome editing: recommendations*, WHO, Geneva 2021, in https://www.who.int/publications-detail-redirect/9789240030381[13-7-2021].

[4] HUMANITY+, "The Transhumanist Declaration", 2009, *Humanity+*, in https://www.humanityplus.org/the-transhumanist-declaration; HUMANITY+, "Transhumanist FAQ 3.0", 2001, *Humanity+*, in https://www.humanityplus.org/transhumanist-faq[2-12-2020]; N. VITA-MORE, "The Transhumanist Manifesto (v.4)", 2020, *Humanity+*, in https://www.humanityplus.org/the-transhumanist-manifesto[11-6-2022].

the situation of the public debate. Chapter 9 is dedicated to the task of refining the predominant values found in the RTA: autonomy, protection, and equality-solidarity. To concretize the ethics of HGE, this book concludes with a series of tables to address this complex topic, along with a synthesis of my personal bioethical analysis on the matter.

Excellence in ethics comprises the skill to do more than just state what should not be done. We need an ethical framework that goes beyond stating "Don't do it", but one that also suggests "What could be done". We need an ethics that inspires innovation so that we can promote the emergence of technologies which are compatible with human flourishing. Humanity will not unlearn how to alter the genome. Bioethics should accompany the progress of gene editing by developing ethical reasoning to inspire ethical decisions by scientists, investors, publishers, communicators, medical personnel, and users in this field. It is far easier to criticize than to provide solutions, but we should at least attempt to do the latter.

Contents

Abbreviations

AI	Artificial Intelligence
ART	Assisted Reproductive Technology
CRISPR	Clustered Regularly Interspaced Short Palindromic Repeats
DIY	Do-It-Yourself
FAQ	Frequently Asked Questions
FHI	Future of Humanity Institute
GG	Global Governance
H+	Humanity+
HE	Human Gene Editing
HHGE	Heritable Human Gene Editing
IVF	In Vitro Fertilization
LDL	Low-Density Lipoproteins
NASEM	National Academies of Sciences, Engineering, and Medicine
RTA	Reflexive Thematic Analysis
SCD	Sickle Cell Disease
STOA	European Parliament's Science and Technology Options Assessment
TALENs	Transcription Activator-Like Effector Nucleases
UDHR	Universal Declaration of Human Rights
UN	United Nations
UNESCO	United Nations Educational, Scientific and Cultural Organization
WHO	World Health Organization
WTA	World Transhumanist Association
ZNFs	Zinc Finger Proteins

Part I

Common Themes Between the Global Governance of Human Genome Editing and Transhumanism

Part I is a synthesis of the published excerpt from my doctoral dissertation in Bioethics: *Common Themes Between Humanity+ Original Documents on Transhumanism and WHO Documents on the Global Governance of Human Genome Editing. Reflexive Thematic Analysis and Implications for Bioethics*, Ateneo Pontificio Regina Apostolorum, Rome 2023.

1 The Global Governance of Human Genome Editing and the World Health Organization Framework

Abstract

This chapter offers an overview of the global governance of human genome editing. It begins by explaining governance as the process of managing common issues. Governance is a complex process because it includes multiple stakeholders and means, beyond governments and regulations. Specifically, the global governance of human genome editing is still in its infancy. This chapter offers a comparative chart regarding relevant texts for the governance of gene editing, including reports, callings, treaties, declarations and statements between 1975 and 2021. Lastly, the three texts published by the WHO in 2021 (the only global framework for the governance of human genome editing to date) are presented.

Keywords

Governance • Human genome editing • World Health Organization

1.1 The Global Governance of Human Genome Editing

Genome editing (also called gene editing or genetic engineering) is used to change an organism's DNA, by adding, removing, or altering genetic material at particular locations in the genome. Advances in gene editing are making these alterations more accurate and efficient. The most common techniques include Zinc Finger Proteins (ZNFs), Transcription Activator-Like Effector Nucleases (TALENs), and Clustered Regularly Interspaced Short Palindromic Repeats (CRISPR). Although CRISPR seems currently to be the cornerstone in genetic engineering, other tools are likely to emerge and improve the editing process. For instance, Prime is a more precise gene editing method that does not require double-strand breaks or donor DNA templates. Prime researchers affirm that this "substantially expands the scope and capabilities of genome editing, and in principle could correct up to

L. Santos, *Governance of Human Gene Editing and Transhumanism*,
Integrated Science 28, https://doi.org/10.1007/978-3-031-70863-3_1

89% of known genetic variants associated with human diseases."[1] Regardless of the specific tool, gene editing is already being used to alter the genome of plants, animals, and humans.[2] The present work focuses exclusively on these applications as they pertain to human cells: human genome editing (HGE).

The Third International Summit on Human Genome Editing, held in London in March 2023, distinguished between three types of HGE: somatic, germline, and heritable HGE.[3] Somatic editing (performed on non-reproductive cells) has proven effective in treating sickle cell disease, and other promising clinical trials for various genetic disorders are in progress. Furthermore, there are ongoing clinical trials for the therapeutic use of somatic editing beyond treating rare genetic disorders, such as for reducing the risk of cardiovascular diseases.[4]

Following Dr He Jiankui's 2018 experiment,[5] in which heritable gene editing was performed and twin babies were born in China, the international scientific community now differentiates *germline HGE* from *heritable* HGE.[6] Germline HGE refers to the editing of human embryos, gametes, or gametes precursor cells in the research setting only, not for human reproduction. Regarding germline HGE, the London 2023 Organising Committee of the Third International Summit on HGE stated that "basic research in this field should continue."[7]

As for heritable HGE, this refers to the gene editing done on human embryos, gametes or their precursor cells to be implanted and used for human reproduction purposes. The 2023 London Organising Committee concluded that heritable HGE "should not be used unless, at a minimum, it meets reasonable standards for safety and efficacy, is legally sanctioned, and has been developed and tested under a system of rigorous oversight that is subject to responsible governance. At this time, these conditions have not been met."[8]

---

[1] Cf. A. V. Anzalone et al., «Search-and-replace genome editing without double-strand breaks or donor DNA», *Nature* 576/7785 (2019), 149–157, in https://www.nature.com/articles/s41586-019-1711-4 [17-6-2021], 149. *Ibid.*

[2] Cf. J. A. Doudna - S. H. Sternberg, *A Crack in Creation: Gene Editing and the Unthinkable Power to Control Evolution*, Houghton Mifflin Harcourt, Boston 2017, xiii–xv.

[3] Cf. The Organising Committee of the Third International Summit on Human Genome Editing, «Statement from the Organising Committee of the Third International Summit on Human Genome Editing», 2023, in https://royalsociety.org/news/2023/03/statement-third-international-summit-human-genome-editing/ [7-3-2023].

[4] Cf. Verve Therapeutics, «Verve Therapeutics Doses First Human with an Investigational In Vivo Base Editing Medicine, VERVE-101, as a Potential Treatment for Heterozygous Familial Hypercholesterolemia», 2022, *Verve Therapeutics*, in https://ir.vervetx.com/news-releases/news-release-details/verve-therapeutics-doses-first-human-investigational-vivo-base/ [3-2-2023].

[5] Cf. M. Marchione, «Chinese researcher claims first gene-edited babies», *AP NEWS* (2018), in https://apnews.com/article/4997bb7aa36c45449b488e19ac83e86d [19-12-2020].

[6] Cf. The Organising Committee of the Third International Summit on Human Genome Editing, «Statement from the Organising Committee of the Third International Summit on Human Genome Editing», 2023, in https://royalsociety.org/news/2023/03/statement-third-international-summit-human-genome-editing/ [7-3-2023].

[7] *Ibid.*, 2.

[8] *Ibid.*

When it comes down to the application of HGE, there are numerous questions that emerge. Even if a type of HGE is possible, should it necessarily be done? Should all possible HGE applications under a category be authorised? Can a transition path be traced out? How can we avoid ethical slippery slopes? How can the differences between countries be considered (from cultural diversity to variations regarding resources and policies)? Who should be the decision-makers? Who will implement these decisions and monitor applications? Which values and principles will be chosen considering our pluralistic world? Will decisions be guided by a moral vision or by public opinion? Which moral vision will enlighten the process? Who will be considered the stakeholders (the majority, minorities, lobbies, directly affected people)? What is to be considered a fruitful public debate? How do the different stakeholders and means come into play? Is a good global management of HGE even possible?

Global governance (GG) is the current framework between international entities to facilitate the management of various global challenges, as a "world government" is neither realistic nor desired. Global governance, for example, is applied to the management of emerging technologies that transcend national borders. The word governance comes from the Greek *κῠ́βερν´ᾰ́ω* or *kubernáō,* which means "*to steer*". The term has been used in numerous texts, for example, *The Governance of England* by Charles Plummer, published in 1885. However, the usage of this word has become increasingly popular in political, social, economic, and ethical discourses since the 1990s.[9]

Today, governance can be defined as "the process of governing, by formal or informal bodies, including governments; in different frameworks, including hierarchy, market, and network; through different measures, including laws, regulations, norms, money, communication, or exchanges; and over different sorts of aspects of collective human life."[10] Governance exists at the regional, national, and international levels. At the international level, the way that various stakeholders exercise different types of power to manage a common issue is called *global governance*. For example, matters such as migration, money laundering, terrorism and bioterrorism, pandemics, loss of biodiversity, climate change, and genocide go beyond the capacities of individual states to solve.[11] These global issues are indeed "problems without passports", as quoted by former UN Secretary General, Kofi Annan.[12]

[9] Cf. H. TEN HAVE (ed.), *Encyclopedia of Global Bioethics*, II, Springer International Publishing, Cham 2016, 1424.
[10] *Ibid.*
[11] Cf. T. G. WEISS, *Global Governance: Why? What? Whither?*, Polity Press, Cambridge 2013, 181.
[12] Cf. *Ibid.*, 180.

To gain a more comprehensive understanding of how governance is far more complex than merely creating regulations and making judicial decisions, a recent example of what occurred in the United States after the 2022 sentence *Dobbs v Jackson*[13] is presented. The Supreme Judges declared that there is no basis for a constitutional right to abortion and that the American Constitution does not prohibit individual states from regulating or prohibiting the procedure. In response to this ruling, some pharmaceutical industries, some employers, businesses, and even the President of the United States, rapidly offered options to pregnant women seeking an abortion. Such efforts varied from easing access to abortive pills,[14] facilitating travel to places where abortion is permitted, and erasing research data that may be related to looking for a termination of pregnancy.[15]

Although this example portrays the complexity of governance at a national level, it helps to better understand how governance is extensively broader than the act of ruling. Factors such as law and enforcement, public opinion, media, private and public sectors, business incentives, patents, insurance, taxes, funding, and many more play a significant role in the process of managing common issues. To exercise governance efficiently, careful attention should be paid to the dynamics of different corporations, organisations, NGOs, religious bodies, media industries, pressure groups, lobbyists, coalitions, civil society actors, activists, and other formal and non-formal role-players.[16]

Bioethics is an interdisciplinary field of study which aims to be a "bridge to the future"[17]. The future is shaped by the stakeholders of global governance. Therefore, exceptional focus should be placed on defining the global governance relating to technologies such as HGE, as it is a tool that powerfully influences the future of humanity. The use of gene editing technologies and its global governance are relevant to current Bioethics.

[13] Cf. «Dobbs v. Jackson Women's Health Organization 597 U.S. ___», 2022, 2228, in https://supreme.justia.com/cases/federal/us/597/19-1392/ [1-2-2023].

[14] Cf. R. ACKERMANNARCHIVE, «Abortion pills via telemedicine: 10 Breakthrough Technologies 2023», 2023, *MIT Technology Review*, in https://www.technologyreview.com/2023/01/09/1064871/abortion-pills-telemedicine-10-breakthrough-technologies-2023/ [3-2-2023].

[15] Cf. «Roe v Wade: Women travelling for abortions will be protected - Biden», *BBC News* (2022), in https://www.bbc.com/news/world-us-canada-62018206 [2-7-2022].

[16] Cf. K. ARAMESH, «Governance», in H. TEN HAVE (ed.), *Encyclopedia of Global Bioethics*, Springer International Publishing, Cham 2016, 1423–1432, 1424.

[17] Cf. V. R. POTTER, *Bioethics: bridge to the future*, Prentice-Hall, Englewood Cliffs, N.J. 1971, in http://catalog.hathitrust.org/api/volumes/oclc/226527.html [12-3-2021].

1.2 Documents on the Governance of Human Genome Editing

Numerous documents address the governance of HGE[18]. They vary in date (from 1975), topic (some focus on all types of human gene editing, some on heritable gene editing, etc.), length (some are brief statements or callings, others are extensive reports) and nature (reports, books, articles, international treaties, international declarations, etc.). The following table outlines information about the main documents considered for this work up to July 2021, prior to the publication of the three texts by the WHO on the global governance of HGE.

Many of the documents presented above call for urgent global governance of HGE. However, neither national bodies nor unions of international bodies could alone provide a framework for the global governance of HGE. It was necessary for a worldwide international body, the World Health Organization (WHO) to develop a framework for global governance of HGE.

In December 2018, the WHO established an Expert Advisory Committee, formed by eighteen experts from different countries and fields of expertise, to advise and make recommendations regarding governance mechanisms for HGE[19]. The Committee worked collaboratively for more than two years, and in July 2021, the WHO published the following documents:

- *Human genome editing: a framework for governance*[20]
- *Human genome editing: recommendations*[21]
- *Human genome editing: position paper.*[22]

As of today, these publications are the only existing documents that articulate a plan for the governance of HGE at a global level. They not only emphasise the

[18] Cf. G. E. Marchant, «Global Governance of Human Genome Editing: What Are the Rules?», *Annual Review of Genomics and Human Genetics* 22/1 (2021), 385–405, in https://www.annualreviews.org/doi/10.1146/annurev-genom-111320-091930 [16-9-2021], 394–396.

[19] More information on the Committee's work, meeting reports, publications and related activities are found at WHO official web page. Cf. who, «Human genome editing», in https://www.who.int/health-topics/human-genome-editing [12-12-2022].

[20] Cf. WHO Expert Advisory Committee on Developing Global Standards for Governance and Oversight of Human Genome Editing, *Human genome editing: a framework for governance*, WHO, Geneva 2021, in https://www.who.int/publications/i/item/9789240030060 [13-7-2021].

[21] Cf. WHO Expert Advisory Committee on Developing Global Standards for Governance and Oversight of Human Genome Editing, *Human genome editing: recommendations*, WHO, Geneva 2021, in https://www.who.int/publications-detail-redirect/9789240030381 [13-7-2021].

[22] Cf. WHO Expert Advisory Committee on Developing Global Standards for Governance and Oversight of Human Genome Editing, *Human genome editing: position paper*, WHO, Geneva 2021, in https://www.who.int/publications-detail-redirect/9789240030404 [13-7-2021].

Table 1.1 Important documents on gene editing before the publication of the WHO framework (2021)

Date	Document	Issued by	Relevance
Jun. 1875	*Summary Statement of the Asilomar Conference on Recombinant DNA molecules*	National Academy of Sciences (USA)	Set forth initial guidelines for recombinant DNA research
Jun. 1997	*Convention on Human Rights and Biomedicine (Oviedo Convention)*, and the recent *Clarifications* (2021)	Council of Europe	The only international legally binding instrument for the protection of human rights in biomedicine Chap. 4—Human genome
Nov. 1997	*Universal Declaration on the Human Genome and Human Rights*	UNESCO	UNESCO country members adopted the declaration unanimously. It refers to the human genome as the "heritage of humanity", and as such, it should be protected and passed on to future generations
Jan. 2015	*A prudent path forward for genomic engineering and germline gene modification*	A group of stakeholders met in California	The goal was to initiate an informed discussion of the uses of genome engineering, and to identify those areas where action is essential to prepare for future developments Conclusions: Discourage germline applications; Forums and information, Transparent research; Globally representative group; Policies
Oct. 2015	*Report of the IBC on updating its reflection on the Human Genome and Human Rights*	International Bioethics Committee (IBC)	Updating ethical reflection is a never-ending commitment. Some revisions of existing declarations should also be considered. The cogency of the principles is untouched. The scope of possible applications is changing and widening every day. Selected areas: direct-to-consumer, precision/personalised medicine, biobanks, and non-invasive pre-natal testing

(continued)

Table 1.1 (continued)

Date	Document	Issued by	Relevance
Dec. 2015	*First International Summit on Human Gene Editing* (Washington, DC)	National Academies of Sciences and Medicine (USA), the Royal Society (UK), and the Chinese Academy of Sciences	First international summit to present and deliberate on the scientific, ethical, legal, social, and governance issues associated with human gene editing.
Sep. 2016	*Genetic Editing: an ethical review*	Nuffield Council on Bioethics (UK)	Issues that should be addressed urgently: the potential application of genome editing in human reproduction and livestock. Issues that may need to be addressed in the near future: gene drives and xenotransplantation. Issues that should be kept under review: governance to develop new cell-based therapies, and to develop new plant strains in agriculture
Feb. 2017	*Human Genome Editing: Science, Ethics, and Governance*	National Academies of Sciences, Engineering, and Medicine (USA)	Recommendations regarding: global principles for research and clinical use; basic research; somatic and germline (heritable) genome editing; enhancement; public engagement.
Jul. 2018	*Genome editing and human reproduction: social and ethical issues*	Nuffield Council on Bioethics (UK)	Prospective parents' interests alongside 2 principles: the welfare of the future person and social justice-solidarity.
Nov. 2018	*Second International Summit on Human Gene Editing* (Honk Kong)	National Academies of Sciences and Medicine (USA), the Royal Society (UK), and the Academy of Sciences of Honk Kong	Shortly after the news about the CRISPR babies. He Jiankui was a speaker. His experiment was considered irresponsible. Topics included the potential benefits and risks of HGE, ethical and cultural perspectives, regulatory and policy considerations, and public engagement efforts.

(continued)

Table 1.1 (continued)

Date	Document	Issued by	Relevance
Mar. 2019	*Adopt a moratorium on heritable genome editing*	Published in *Nature*. 18 signatories (scientists and ethicists from 7 countries, including Emmanuelle Charpentier, co-creator of CRISPR-Cas9)	Call for a global moratorium on changing heritable DNA. Call for the establishment of an international framework in which nations voluntarily commit to not approving any use of clinical germline editing unless certain conditions are met.
Jul. 2019	*Statement on governance and oversight of human genome editing*	WHO	WHO Director-general stated that "it would be irresponsible at this time for anyone to proceed with clinical applications of human germline genome editing. (…) I have accepted the interim recommendations of WHO's Expert Advisory Committee that regulatory authorities in all countries should not allow any further work in this area until its implications have been properly considered."[23]
Sep. 2020	*Heritable Human Genome Editing*	National Academies of Sciences, Engineering, and Medicine (USA)	Translational pathway for clinical use of heritable HGE Preclinical and clinical requirements Recommendations on requirements for oversight frameworks
May 2021	*Guidelines for Stem Cell Research and Clinical Translation*	International Society for Stem Cell Research (ISSCR)	Recommendations regarding heritable HGE (not for reproduction at this time; evaluate case-by-case in research; regulatory and ethical framework needed; safety-ethical-societal issues; risk vs benefit; follow-up)

need for global governance of HGE, but also present a framework that considers the multiple stakeholders and tools involved. Furthermore, the WHO Committee

[23] WHO, «Statement on governance and oversight of human genome editing», 2019, in https://www.who.int/news/item/26-07-2019-statement-on-governance-and-oversight-of-human-genome-editing [13-7-2021].

incorporated various suggestions made by the documents mentioned in Table 1.1. These three WHO publications are the answer to the international call for global governance of HGE and constitute the most pertinent texts regarding the global governance of HGE to date. It is important to become familiar with the structure and contents of these documents, which will be key to the Reflexive Thematic Analysis (Chap. 3).

1.2.1 Human Genome Editing: A Framework for Governance[24]

The content of WHO framework for global governance of HGE is divided into:

- *Foreword, Acknowledgements, Executive summary*
- *Part 1. Introduction*
- *Part 2. Good governance of new and emerging technologies*
- *Part 3. Governance of human genome editing*
  Special challenges: postnatal somatic, prenatal (in utero) somatic, heritable, human epigenetic editing, enhancement.
- *Part 4. Tools, institutions and processes for governance of human genome editing*
  Declarations, treaties, conventions, legislation and regulations, Judicial rulings, Ministerial decrees. Conditions on research funding, Moratoria, Accreditation, registration or licensing. National science and medicine societies and institutions. Patents and licences/Professional self-regulation. Public advocacy and activism. Research ethics guidelines and research ethics review. Collaboration with publishers and conference organisers. Education and training of researchers and clinicians.
- *Part 5. Scenarios*
  Somatic human genome editing: clinical trials for sickle-cell disease, clinical trials for Huntington disease, unscrupulous entrepreneurs and clinics, to enhance athletic ability. Heritable human genome editing (for reproduction), unscrupulous entrepreneurs and clinics expanding assisted reproduction. Prenatal (in utero) somatic human genome editing: clinical trials for cystic fibrosis.
- *Part 6. Implementation, metrics and review*
  Implementation of the governance framework and associated measures. Metrics. Reviewing and updating the governance framework.
- *Annex. Meetings, consultations and webinars: participants*

[24] WHO Expert Advisory Committee on Developing Global Standards for Governance and Oversight of Human Genome Editing, *Human genome editing: a framework for governance*, WHO, Geneva 2021, in https://www.who.int/publications/i/item/9789240030060 [13-7-2021].

1.2.2 Human Genome Editing: Recommendations[25]

This second WHO document focuses on the recommendations for good global governance of HGE. While the document regarding the framework focused on the challenges, scenarios, tools, and processes for governance, this document stated the recommended actions.

Its content is divided into the following parts:

- *Foreword, Acknowledgements, Executive summary*
- *Part 1. Background*
- *Part 2. Recommendations:*
  - Leadership by the WHO and its Director-General
  - International collaboration for effective governance and oversight
  - Human genome editing registries
  - International research and medical travel
  - Illegal, unregistered, unethical or unsafe research and other activities
  - Intellectual property
  - Education, engagement and empowerment
  - Ethical values and principles for use by WHO
  - Review of the recommendation
- *Annex. Meetings, consultations and webinars: participants*

1.2.3 Human Genome Editing: Position Paper[26]

This final document repeats the *Acknowledgements, Executive summary*, and *Background* information present in the document *Human Genome Editing: A Framework for Governance.* It also synthesises the recommendations given in *Human Genome Editing: recommendations.*

The document is divided into the following sections:

- *Acknowledgements, Executive summary, Background information*
- *Preventing premature use of human genome editing*
- *Governance framework*
- *Recommendations*

[25] WHO Expert Advisory Committee on Developing Global Standards for Governance and Oversight of Human Genome Editing, *Human genome editing: recommendations*, WHO, Geneva 2021, in https://www.who.int/publications-detail-redirect/9789240030381 [13-7-2021].

[26] WHO Expert Advisory Committee on Developing Global Standards for Governance and Oversight of Human Genome Editing, *Human genome editing: position paper*, WHO, Geneva 2021, in https://www.who.int/publications-detail-redirect/9789240030404 [13-7-2021].

To conclude, the World Health Organization recognises that gene editing extends beyond national borders, and therefore global action is necessary.[27] Within the framework, the WHO lists twelve different sets of *tools, institutions and* processes that could potentially be involved in the governance of HGE. It also includes recommendations relating to laws and regulations, patents and licenses, research funding, professional self-regulation, the role of professional bodies, publishers, public advocacy and activism.[28] The framework presents a set of principles to be considered to inform *how* decisions are made: *Openness, Transparency, Honesty and Accountability, Responsible regulatory stewardship, Responsible stewardship of science, and Responsible stewardship of research resources.* The WHO framework also presents a set of principles to inform *what* decisions are made: *Inclusiveness, Caution, Fairness, Social justice, Non-discrimination, Equal moral worth, Respect for persons, Solidarity, and Global health justice.*[29]

It is important to note that the three texts are not the global governance of HGE in itself, as the process of governing involves many other agents apart from the WHO. As one of the titles displays, the publications offer the *framework* recommended by the WHO for the global governance of HGE.

[27] Cf. WHO Expert Advisory Committee on Developing Global Standards for Governance and Oversight of Human Genome Editing, *Human genome editing: a framework for governance*, WHO, Geneva 2021, in https://www.who.int/publications/i/item/9789240030060 [13-7-2021], x.

[28] Cf. *Ibid.*, x–xi.

[29] Cf. *Ibid.*, 13–14.

Transhumanism and Its Original Documents

2

Abstract

This chapter offers an overview of Transhumanism. It synthesises the origins of the word Transhumanism and its contemporary meaning. To better understand this intellectual and cultural movement, this chapter portrays some of the most influential people for Transhumanism, with information about their lives, education, importance within Transhumanism, and examples of works. Transhumanism is a vast and largely defined movement, so the currents within Transhumanism are also presented. Lastly, the chapter gives an overview of three so-called "original documents on Transhumanism", according to Humanity+ (the former World Transhumanist Association). *The Transhumanist Declaration*, the *Transhumanist FAQ 3.0* and *The Transhumanist Manifesto v.4* contain essential convergences among the proponents of this movement.

Keywords

Transhumanism • Technology • Enhancement • Gene editing

L. Santos, *Governance of Human Gene Editing and Transhumanism*,
Integrated Science 28, https://doi.org/10.1007/978-3-031-70863-3_2

2.1 Transhumanism

2.1.1 Introduction to Transhumanism

The transhumanist movement is not simple to define. The word "transhuman" is employed to describe several concepts at the same time. Also, as leading transhumanist Bostrom[1] states, Transhumanism is not "a philosophy with a fixed set of dogmas (…) The ideology is meant to evolve and be reshaped as we move along."[2]

The term *trasumanar* was coined by Dante Alighieri, in the *Divine Comedy*, before entering the Paradise: "trasumanar significar per verba/non si porìa: però l'essemplo basti/a cui esperienza grazia serba".[3] In the text, Dante's neologism *trasumanar* meant going beyond the limits of human nature, which is transformed to adhere to the divine nature and cannot be described.

Transhumanism in its current sense is often linked to Julian Huxley, an evolutionary biologist and the first Director-General of UNESCO. Julian had two famous brothers: Biologist Andrew Huxley, and Aldous Huxley, the author of *Brave New World*. The Huxley brothers were grandchildren of Thomas Henry Huxley, well known as "Darwin's bulldog" a nickname which stemmed from his public support for his friend's evolutionary ideas.[4] Julian Huxley popularised the word Transhumanism in his book *New Bottles For New Wine*, where he wrote:

> The human species can, if it wishes, transcend itself - not just sporadically, an individual here in one way, an individual there in another way, but in its entirety, as humanity. We need a name for this new belief. Perhaps transhumanism will serve: man remaining man, but transcending himself, by realizing new possibilities of and for his human nature.[5]

---

[1] Dr. Nick Bostrom is a Swedish-born philosopher with a background in theoretical physics, computational neuroscience, logic, and artificial intelligence, as well as philosophy. A founder of the World Transhumanist Association (WTA), now Humanity+, he is the author of numerous publications. Until 2024 he was a Professor at Oxford University, where he lead the Future of Humanity Institute (recently closed) as its founding director. Cf. N. BOSTROM, «Nick Bostrom's Home Page», in https://www.nickbostrom.com/ [12-7-2020].

[2] N. BOSTROM, «What is transhumanism?», 2001, in https://www.nickbostrom.com/old/transhumanism.html [15-9-2020].

[3] D. ALIGHIERI, *La Divina Commedia*, 1320, in https://divinacommedia.weebly.com/ [7-11-2022] v.70.

[4] Cf. A. J. DESMOND, «Thomas Henry Huxley», in *Encyclopedia Britannica*, Philosophy & Religion, Encyclopedia Britannica, Chicago 2021, in https://www.britannica.com/biography/Thomas-Henry-Huxley [20-4-2021].

[5] JULIAN HUXLEY, *New Bottles For New Wine*, Chatto and Windus LTD., London 1957, in http://archive.org/details/NewBottlesForNewWine [15-5-2020], 17.

Bostrom reported that the word *transhumanism* appeared to have been first used by Julian Huxley in his book *New Bottles for New Wine*, published in 1957.[6] However, it was reported that the word was used at a conference in 1939 by a French polytechnician, Jean Coutrot, a friend of Julian's brother, Aldous Huxley.[7] Subsequently, Julian used the term *transhumanism* during a lecture in 1951, and then in his publication in 1957.[8] Therefore, it is more correct to say that Julian Huxley popularised the term *transhumanism*.

Humanity+ (H+), formerly the World Transhumanist Association (WTA), presents Transhumanism as "the intellectual and cultural movement that affirms the possibility and desirability of fundamentally improving the human condition through applied reason, especially by developing and making widely available technologies to eliminate aging and to greatly enhance human intellectual, physical, and psychological capacities."[9]

Humanity+ emphasises that the objective is not to improve the conditions of human life but to improve the human condition. According to the *Transhumanist FAQ 3.0*, "an improvement to the human condition is a change that gives increased opportunity for individuals to shape themselves and their lives according to their informed wishes."[10] What type of changes are they implying? The enhancement proposed by transhumanists is not the physical, cultural, or moral improvement achieved by traditional methods. Technology is the key to radical enhancement and to moving beyond what is currently considered human. According to transhumanists, we are "not limited to traditional humanistic methods, such as education and cultural development. We can also use technological means that will eventually enable us to move beyond what some would think of as "human".[11]

Regarding Transhumanism as a cultural movement, James Hughes, a former Executive Director of the World Transhumanist Association, stated that:

> We need transhumanist think tanks, journals, conferences and lobbyists. We need transhumanists meeting the bioLuddites toe-to-toe in the public square, defending the rights of persons to use reason to control their own affairs. We need transhumanist clubs and study

---

[6] Cf. N. BOSTROM, «A History of Transhumanist Thought», in M. RECTENWALD - L. CARL (edd.), Academic Writing Across the Disciplines, Pearson Longman, New York 2011, 7.

[7] Cf. «Coutrot, Jean», 2019, *Sciences Po Centre d'histoire*, in https://www.sciencespo.fr/histoire/fr/fonds-archive/coutrot-jean.html [15-11-2022]; Cf. N. BOSTROM, «A History of Transhumanist Thought», 7.

[8] Cf. C. BYK, «Transhumanism: from Julian Huxley to UNESCO: What Objective for International Action?», *Jahr - European Journal of Bioethics* 12/1 (2021), 141–162, in https://www.jahr-bioethics-journal.com/index.php/JAHR/article/view/565 [15-11-2022], 143; Cf. N. BOSTROM, «A History of Transhumanist Thought», in M. RECTENWALD - L. CARL (edd.), Academic Writing Across the Disciplines, Pearson Longman, New York 2011, 7.

[9] HUMANITY+, «Transhumanist FAQ 3.0», 2001, *Humanity+*, in https://www.humanityplus.org/transhumanist-faq [2-12-2020].

[10] *Ibid.*

[11] *Ibid.*

groups on the campuses, and in every city in every country, educating the public about the threats and promises to come.[12]

In a recent publication, Coenen stated that: "the transhumanist worldview is evidently not merely being promoted by transhumanist organizations, a few academics (working in applied ethics, for example) and isolated Political activities (such as the NBIC initiative), but also by key figures and companies in the IT industry that is so fundamental for our society today."[13] Indeed, Transhumanism can be observed not only within the culture of science fiction (from movies such as *Her*, *Trascendence*, *In Time*, *Ex Machina*, *Blade Runner*, *Black Mirror*, and *Upload),* but also within celebrities such as Elon Musk (inventor of Uber and SpaceX, and intellectually influenced by transhumanist Nick Bostrom) and Ray Kurzweil (Engineer at Google).[14]

The statement "The transhumanists are not content to simply discuss the merit of enhancement, rather they are working to build a world favorably aligned."[15] is consistent with the affirmation that transhumanism is an intellectual *and* cultural movement.[16] As one can see, transhumanism is acting on a variety of different fronts to reach its goals.

For transhumanists, what is considered good is dependent on what we decide human beings could, or should, become. This trend poses a major challenge for Global Bioethics. The European Group on Ethics of Science and New Technologies explains: "Just as we started to find some relatively firm foundations for ethics in our common humanity, our evolutionary biology, psychology and common history, some suggest that we consider humanity as something that is not given and can be overcome and transcended by technological design and engineering."[17] The European Group also notes that this shift in the anthropological scenario brings

[12] J. J. HUGHES, *Citizen Cyborg: Why Democratic Societies Must Respond To The Redesigned Human Of The Future*, Basic Books, Cambridge 2004, 260.

[13] C. COENEN, «Transcending Natural Limitations: The Military-Industrial Complex and the Transhumanist Temptation», in W. HOFKIRCHNER - H.-J. KREOWSKI (edd.), *Transhumanism: The Proper Guide to a Posthuman Condition or a Dangerous Idea?*, Cognitive Technologies, Springer International Publishing, Cham 2021, 97–110, 101.

[14] Cf. M. A. CRESPO- RODRÍGUEZ, *Zoltan Istvan y el Partido Transhumanista: Política y transhumanismo en el siglo XXI*, University of Puerto Rico, Rio Piedras 2018, in https://www.proquest.com/docview/2023807738/abstract/B0C097A17A8F452BPQ/1 [12-5-2021], 12.

[15] S. LILLEY, *Transhumanism and Society: The Social Debate over Human Enhancement*, SpringerBriefs in Philosophy, Springer, Heidelberg 2013, in https://www.springer.com/it/book/9789400749801 [20-5-2021], 4.

[16] Cf. HUMANITY+, «Transhumanist FAQ 3.0», 2001, *Humanity+*, in https://www.humanityplus.org/transhumanist-faq [2-12-2020].

[17] EUROPEAN GROUP ON ETHICS IN SCIENCE AND NEW TECHNOLOGIES, *Values for the Future: The Role of Ethics in European and Global Governance*, Publications Office of the European Union, Brussels 2021, in https://ec.europa.eu/info/files/values-future-role-ethics-european-and-global-governance_en [16-11-2021], 13.

forth consequences for ethics: "What we are, and what is good for us human beings, then depends on what we decide human beings could or should become."[18]

Transhumanist ideas such as discontentment with nature, materialism, malleability, directed evolution, and safe enhancement for all are likely to influence stakeholders' decisions regarding HGE. Furthermore, transhumanist reasonings, proposals, and actions add a new nuance to global governance technologies such as HGE. Indeed, the WHO Committee held a webinar on June 11th, 2020, asking for biohackers', DIY community labs', and transhumanists' perspectives on HGE. Leading transhumanist Nick Bostrom was among the consulted people.[19] Transhumanism is already a relevant topic in Bioethics, even at the highest levels such as the WHO, especially when working on the global governance of new technologies for human health. To be clear, relevance does not mean that the global governance of HGE is aligned with Transhumanism. However, it indicates the importance of not ignoring the transhumanist movement.

2.1.2 Key Individuals for Transhumanism

It is always an enlightening perspective to note the people behind a particular movement. Within this section, some important figures within the Transhumanism movement are considered. Certain individuals are more influential in the academic field, whilst others offer a more cultural, social, or political impact. The order of appearance is arbitrary and does not intend to show a hierarchy amongst the important individuals for Transhumanism. Furthermore, this table is not meant to be exhaustive, as many other individuals could also be considered important contributors to Transhumanism.

Follows twelve of the most influential people for Transhumanism, with information about their lives, education, importance within the transhumanist movement, and examples of publications related to the topic of this book.[20]

[18] *Ibid.*

[19] Cf. WHO Expert Advisory Committee on Developing Global Standards for Governance and Oversight of Human Genome Editing, *Human genome editing: a framework for governance*, WHO, Geneva 2021, in https://www.who.int/publications/i/item/9789240030060 [13-7-2021], 78–79.

[20] More information about important people for Transhumanism can be found in the section titled *Who's Who of Transhumanism*. Cf. R. MANZOCCO, *Transhumanism - Engineering the Human Condition: History, Philosophy and Current Status*, Springer, New York 2019, 61–72.

Table 2.1 Important figures for transhumanism—Huxley

Sir Julian Sorell Huxley	
Keywords	• Word *Transhumanism*
Personal Life	• 1887–1975, UK • Brother of another famous biologist, Andrew Huxley, and of Aldous Huxley, the author of Brave New World • Grandson of Thomas Henry Huxley, known as the "Darwin's bulldog" for the public advocacy for his friend's evolutionary ideas
Education Background	• Biology (Balliol College)
Importance for Transhumanism	• He popularised the word Transhumanism • Secretary of the Zoological Society of London (1935–1942) • First Director-General of UNESCO (1946–1948) • Founding member of the World Wildlife Fund (WWF) • President of the British Eugenics Society (1959–1962) • First President of the British Humanist Association
Main Pertinent Publications	• *New Bottles For New Wine* • *Transhumanism* • *Religion without Revelation* • *Evolution, The Definitive Edition: The Modern Synthesis*

Table 2.2 Important figures for transhumanism—More

Max-More (born Max T. O'Connor)	
Keywords	• Extropy • Proactionary Principle
Personal Life	• 1964, UK- • Spouse: Natasha Vita-More
Education Background	• Philosophy (University of Southern California) • Philosophy, Politics, and Economics (Oxford University)
Importance for Transhumanism	• Co-founder of Extropy Institute, an organisation crucial in building the transhumanist movement since 1990 • Ambassador and President Emeritus of Alcor Life Extension (cryonics)
Main Pertinent Publications	• *The Proactionary Principle* • *A Letter to Mother Nature: Amendments to the Human Constitution* • *Philosophy of Transhumanism*

Table 2.3 Important figures for transhumanism—Vita-More

Natasha Vita-More (born Natasha Vita)	
Keywords	• H+ • Manifesto
Personal Life	• 1950, US- • Spouse: Max More
Education Background	• Media Arts, Design and Architecture • Communications and Media Studies (University of Plymouth) • Sciences and Technology - Future Studies (University of Houston)
Importance for Transhumanism	• Strategic designer, author, speaker, and innovator within the scientific and technological framework of human enhancement and life extension • Humanity+ Executive Director • Founder of The Center for Transhumanist Studies (CTS) • Founder of Humanity+ Decentralized Autonomous Organization (H+DAO) • Senior Fellow at the Institute for Ethics and Emerging Technologies (IEET)
Main Pertinent Publications	• *The Transhumanist Manifesto* • *The Transhumanist Reader: Classical and Contemporary Essays on the Science, Technology, and Philosophy of the Human Future* • *Transhumanism: What is it?*

Table 2.4 Important figures for transhumanism—Pearce

David Pearce	
Keywords	• Hedonist Imperative
Personal Life	• 1959, UK- • A vegan advocate. His parents, grandparents and three of his great-grandparents were all vegetarian. From a young age, he was concerned with death and ageing, the problem of suffering
Education Background	• Scholarship to study Politics, Philosophy and Economics at Oxford • Never finished his degree
Importance for Transhumanism	• Co-founder of the World Transhumanist Association (WTA), now H+, with Bostrom • Senior Fellow at the Institute for Ethics and Emerging Technologies (IEET)
Main Pertinent Publications	• *The Hedonist Imperative* • *The Biointelligence Explosion* • *Can Biotechnology Abolish Suffering?*

Table 2.5 Important figures for transhumanism—Hughes

James Hughes	
Keywords	• WTA, Democratic Transhumanism
Personal Life	• 1961, US- • Temporarily ordained as a Buddhist monk in 1984 in Sri Lanka
Education Background	• Sociology (University of Chicago)
Importance for Transhumanism	• Professor at Trinity College, Connecticut • Co-founder with Bostrom and now Executive director of the Institute for Ethics and Emerging Technologies (IEET) • Executive director of the World Transhumanist Association (WTA) from 2004 to 2006
Main Pertinent Publications	• *Democratic Transhumanism* • *Citizen Cyborg* • *The Politics of Transhumanism and the Techno-Millennial Imagination, 1626–2030*

Table 2.6 Important figures for transhumanism—FM-2030

FM-2030 (born Fereydoon M. Esfandiary)	
Keywords	• Name • Cryonics
Personal Life	• 1930, Belgium—2000, US • Iranian-American • Son of an Iranian diplomat, he lived in 17 countries in his first 11 years • Olympics athlete (1948) • United Nations diplomat (1952–1954)
Education Background	• University of California
Importance for Transhumanism	• Author, philosopher, designer, long-range planner, lecturer, and transhumanist pioneer • He changed his name to break naming conventions based on collectivism and to express his hope he would celebrate his 100th birthday in 2030 • Dead with pancreatic cancer, FM-2030 was the first person to be vitrified, rather than simply frozen as previous cryonics patients had been
Main Pertinent Publications	• *Are you a Transhuman?: Monitoring and Stimulating Your Personal Rate of Growth in a Rapidly Changing World* • *Optimism One: The Emerging Radicalism*

Table 2.7 Important figures for transhumanism—Kurzweil

Ray Kurzweil	
Keywords	• Singularity
Personal Life	• 1948, US-
Education Background	• Computer Sciences (Massachusetts Institute of Technology) • 21 honorary doctorates
Importance for Transhumanism	• Google Director of Engineering with a focus on AI • Public advocate for Transhumanism • With the collaboration of Google and NASA, he founded the Singularity University
Main Pertinent Publications	• *The Singularity is Near: When Humans Transcend Biology* • *The Age of Spiritual Machines: When Computers Exceed Human Intelligence* • *How to Create a Mind: The Secret of Human Thought Revealed*

Table 2.8 Important figures for transhumanism—Rothblatt

Martine Rothblatt (born Martin Rothblatt)	
Keywords	• Transgender-Transhuman • Innovation
Personal Life	• 1954, US- • Spouse: Bina Rothblatt • 1994: Rothblatt's daughter was diagnosed with a life-threatening disease • 1994: sex reassignment surgery
Education Background	• Communication Studies (University of California) • Medical ethics (Barts and The London School of Medicine and Dentistry)
Importance for Transhumanism	• Author, entrepreneur, and transgender rights advocate • Forbes #44 on the 2022 America's Self-Made Women • Founder and CEO of United Therapeutics • Executive Advisor at H+ • Senior Fellow at the Institute for Ethics and Emerging Technologies (IEET) • Founder of Terasem Movement, which promotes cryogenics, biotechnology, and cyber consciousness
Main Pertinent Publications	• *From Transgender to Transhuman: A Manifesto On the Freedom Of Form* • *Virtually Human*

Table 2.9 Important figures for transhumanism—Istvan

Zoltan Istvan	
Keywords	• US Transhumanist Party
Personal Life	• 1973, US-
Education Background	• Philosophy and Religion (Columbia University)
Importance for Transhumanism	• Journalist and political candidate • Founder of the U.S. Transhumanist Party • 2016: ran for President • 2017: ran for Governor of California • 2018: ran for election as a member of the Libertarian Party. • 2019: ran for the Republican Party's nomination in the 2020 elections against Trump
Main Pertinent Publications	• *Upgrading America: The Political Writings of Zoltan Istvan* • *The Anti-Deathist: Writings of a Radical Longevity Activist* • *Philosophy and the Future: A Transhumanist Examination of Where We're Going*

Table 2.10 Important figures for transhumanism—Bostrom

Nick Bostrom	
Keywords	• WTA • Enhancement • Existential Risks
Personal Life	• 1973, Sweden-
Education Background	• Philosophy (London School of Economics) • Computational Neuroscience (King's College London) • Philosophy, Mathematics, Mathematical Logic, and Artificial Intelligence (University of Gothenburg) • Philosophy and Physics (Stockholm University)
Importance for Transhumanism	• Former Professor at the University of Oxford and Founding director of the Future of Humanity Institute (FHI), closed in 2024 • Co-founder of the World Transhumanist Association (WTA), now H+, with Pearce • Co-founder of the Institute for Ethics and Emerging Technologies (IEET) with Hughes • Work on existential risk, the anthropic principle, human enhancement ethics, and superintelligence
Main Pertinent Publications	• *Human Genetic Enhancements: A Transhumanist Perspective* • *Dignity and Enhancement* • *The Future of Humanity* • *Global Catastrophic Risks* • *Human Enhancement Ethics: The State of the Debate*

Table 2.11 Important figures for transhumanism—Sandberg

Anders Sandberg	
Keywords	• Morphological Freedom
Personal Life	• 1972, Sweden-
Education Background	• Computational Neuroscience (Stockholm University)
Importance for Transhumanism	• Former Senior Research Fellow at the Future of Humanity Institute (FHI) at the University of Oxford
Main Pertinent Publications	• *Morphological Freedom—Why We Not Just Want It, but Need It* • *The Wisdom of Nature: An Evolutionary Heuristic for Human Enhancement* • *Cognitive Enhancement: Methods, Ethics, Regulatory Challenges* • *Who Should We Fear More: Biohackers, Disgruntled Postdocs, or Bad Governments? A Simple Risk Chain Model of Biorisk*

Table 2.12 Important figures for transhumanism—Savulescu

Julian Savulescu	
Keywords	• Procreative Beneficence
Personal Life	• 1963, Australia-
Education Background	• Medicine and Neuroscience (Monash University) • Thesis supervisor: Peter Singer
Importance for Transhumanism	• Founder and former director of Uehiro Centre for Practical Ethics at the University of Oxford • Former editor and current board member of the Journal of Medical Ethics
Main Pertinent Publications	• *Procreative beneficence: why we should select the best children* • *Human Enhancement* • *The Moral Obligation to Create Children with the Best Chance of the Best Life* • *An ethical pathway for gene editing* • *Moral reasons to edit the human genome: picking up from the Nuffield report* • *Unfit for the Future: The Need for Moral Enhancement* • *Is Gene Editing Harmless? Two Arguments for Gene Editing*

The above information about influential people for Transhumanism helps us to learn more about their lives and work. Through them, more can be grasped about what Transhumanism is. The tables also reveal some commonalities:

- *Date.* The majority of the influential figures for Transhumanism were born in the mid-20th century.
- *Countries.* The majority of these individuals were born or worked in Europe (especially in the United Kingdom and Sweden) or the United States.
- *Gender.* Except for Natasha Vita-More, all of the figures listed in the table were born biologically male (including transgender Martine Rothblatt).

2.1.3 Currents within Transhumanism

Naturally, there are diverse perspectives among transhumanist proponents. The terms below presents a summary of existing currents within Transhumanism:

- *Extropy.* it emphasises the Seven Extropian Principles (perpetual progress, self-transformation, practical optimism, intelligent technology, open society, self-direction, and rational thinking). These are meant to codify general attitudes rather than specific dogmas.[21] Max More is the primary thinker for this perspective.
- *Democratic transhumanism.* it advocates for the right to use technology in order to transcend the limitations and the extension of democratic concerns. Supporters of this perspective aim to go beyond *legal* equality and liberty into *economic and cultural* equality and liberty, to ensure the protection of values such as equality, solidarity, and democratic participation in a transhuman context.[22] Hughes is the primary thinker for this perspective.
- *The Hedonistic Imperative.* it advocates for "paradise-engineering" and for a biological program that eliminates all forms of cruelty, suffering, and malaise. From a short-term perspective, an individual's emotional well-being may be enriched by mood-enhancing drugs. From a long-term perspective, supporters of this vision claim that it will be feasible to rewrite the vertebrate genome, redesign the global ecosystem, and use biotechnology to completely abolish suffering throughout the living world.[23] Pearce is the primary thinker for this perspective.
- *Singularitarianism.* it focuses on technologies that can potentially lead to the rise of smarter-than-human intelligence, such as brain-computer interfacing and AI. In the present, human intelligence is ultimately the source of our

[21] Cf. HUMANITY+, «Transhumanist FAQ 3.0», 2001, *Humanity+*, in https://www.humanityplus.org/transhumanist-faq [2-12-2020].
[22] Cf. *Ibid.*
[23] Cf. *Ibid.*

technology. Singularitarians expect the technological creation of smarter-than-human intelligence to be a watershed moment in history.[24] Kurzweil is a strong representative of this perspective.

- *Theoretical transhumanism*: this is not so much a particular method for conducting Transhumanism as a research topic, but instead is the study of the constraints, possibilities, and consequences of potential future trajectories of technological and human development. This perspective explores the theoretical tools of economics, game theory, evolution theory, probability theory, and theoretical applied science. It also investigates the ethical issues related to the transhumanist project.[25] Bostrom is a strong representative of Theoretical Transhumanism.
- *Salon transhumanism*: this is a network of people who share similar interests and enjoy spending long hours conversing about transhumanist matters, either virtually or through face-to-face interactions.[26]
- *Transhumanism in arts and culture*: Transhumanism as a source of inspiration for artistic creation and cultural activities, including the effort to communicate transhumanist ideas and values to a broader audience.[27] Vita-More is the representative of this trend.

2.2 Documents on Transhumanism

In addition to the main publications of transhumanists presented in Tables 2.1, 2.2, 2.3, 2.4, 2.5. 2.6, 2.7, 2.8, 2.9, 2.10, 2.11 and 2.12, there are various other significant works on Transhumanism, such as *The Transhumanist Declaration,*[28] *Philosophy of Transhumanism*[29] *A History of Transhumanist Thought,*[30] *History of Transhumanism*[31] *Transhumanist FAQ,*[32] *What is transhumanism?,*[33] *In Defense of*

[24] Cf. *Ibid.*

[25] Cf. *Ibid.*

[26] Cf. *Ibid.*

[27] Cf. *Ibid.*

[28] Cf. HUMANITY+, «The Transhumanist Declaration», 2009, *Humanity+*, in https://www.humanityplus.org/the-transhumanist-declaration [11-6-2022].

[29] Cf. M. MORE, «Philosophy of Transhumanism», *Humanity+*, in https://www.humanityplus.org/philsophy-of-transhumanism [19-11-2022].

[30] Cf. N. BOSTROM, «A History of Transhumanist Thought», in M. RECTENWALD - L. CARL (edd.), Academic Writing Across the Disciplines, Pearson Longman, New York 2011.

[31] Cf. N. VITA- MORE, «History of Transhumanism», in N. LEE (ed.), *The Transhumanism Handbook*, Springer International Publishing, Cham 2019, 49–61.

[32] Cf. HUMANITY+, «Transhumanist FAQ 3.0», 2001, *Humanity+*, in https://www.humanityplus.org/transhumanist-faq [2-12-2020].

[33] Cf. N. BOSTROM, «What is transhumanism?», 2001, in https://www.nickbostrom.com/old/transhumanism.html [15-9-2020].

Posthuman Dignity,[34] *Transhumanist Ethics,*[35] *Transhumanist Values,*[36] *Transhumanism: The World's Most Dangerous Idea?,*[37] *The Transhumanism Handbook,*[38] *The Transhumanist Reader: Classical and Contemporary Essays on the Science, Technology, and Philosophy of the Human Future,*[39] *Mutare o Perire,*[40] *Designer Evolution: A Transhumanist Manifesto,*[41] *The Transhumanist Manifesto,*[42] *H+/-: Transhumanism and Its Critics,*[43] *Post- and Transhumanism.*[44]

Is there any text or set of texts stating shared ideas among transhumanists? In 2008, the World Transhumanist Association (WTA), founded in 1998 by Bostrom and Pearce, rebranded itself as Humanity+ (H+). Humanity+ defines itself as a non-profit international educational organisation that advocates for the ethical use of emerging technologies and empirical science to enhance present human capacities.[45] The Executive Director of H+ is a widely known transhumanist, Natasha Vita-More. The current Board of Directors includes individuals such as Ben Goertzel, José Cordeiro, Nell Watson, Amy Li, Paul Spiegel, Gabriel Rothblatt (the son of Martine Rothblatt) and Dier Coeurnelle. Furthermore, H+'s Executive Advisors comprises various transhumanist figures including Martine Rothblatt, Max More, Audrey de Grey, Susan Schneider, Anders Sandberg and many more.[46]

---

[34] Cf. N. Bostrom, «In Defense of Posthuman Dignity», *Bioethics* 19/3 (2005), 202–214, in https://onlinelibrary.wiley.com/doi/abs/10.1111/j.1467-8519.2005.00437.x [10-4-2021].

[35] Cf. N. Bostrom, «Transhumanist Ethics», in https://nickbostrom.com/ethics/transhumanist.pdf [10-6-2020].

[36] Cf. N. Bostrom, «Transhumanist Values», *Journal of Philosophical Research* 30/9999 (2005), 3–14.

[37] Cf. N. Bostrom, «Transhumanism: The World's Most Dangerous Idea?», 2004, in https://www.nickbostrom.com/papers/dangerous.html [7-10-2020].

[38] Cf. N. Lee (ed.), *The Transhumanism Handbook*, Springer International Publishing, Cham 2019.

[39] Cf. M. More - N. Vita- More (edd.), *The Transhumanist Reader*, The Transhumanist Reader: Classical and Contemporary Essays on the Science, Technology, and Philosophy of the Human Future, Wiley-Blackwell, Chichester 2013, in https://doi.org/10.1002/9781118555927.ch5 [14-4-2020].

[40] R. Campa, *Mutare o perire. La sfida del transumanesimo*, Sestante Edizioni, Bergamo 2010, 18.

[41] Cf. S. Young, *Designer Evolution: A Transhumanist Manifesto*, Prometheus, Amherst 2005.

[42] Cf. N. Vita- More, «The Transhumanist Manifesto (v.4)», 2020, *Humanity+*, in https://www.humanityplus.org/the-transhumanist-manifesto [11-6-2022].

[43] Cf. G. R. Hansell - W. Grassie (edd.), *H+/-: Transhumanism and Its Critics*, Metanexus Institute, Philadelphia, PA 2011.

[44] Cf. R. Ranisch - S. L. Sorgner (edd.), *Post- and Transhumanism*, Beyond Humanism: Trans-and Posthumanism, Peter Lang, Frankfurt am Main, Berlin, Bern, Bruxelles, New York, Oxford, Wien 2015, in https://doi.org/10.3726/978-3-653-05076-9 [23-3-2021].

[45] Cf. Humanity+, «Mission», *Humanity+*, in https://www.humanityplus.org/about [11-6-2022].

[46] Cf. Humanity+, «Leadership», *Humanity+*, in https://www.humanityplus.org/leadership [14-10-2022].

The official website of H+ (July 2023) calls three particular texts the "Original Documents on Transhumanism"[47]: *The Transhumanist Declaration,*[48] *Transhumanist FAQ 3.0,*[49] and *The Transhumanist Manifesto.*[50] By referring to these three texts as the "Original Documents on Transhumanism", Humanity+ gives them priority over other transhumanist works. Although H+ does not explicitly state why it has chosen these three texts, upon reading them, one can instinctively sense their elevated importance for Transhumanism. It is important to become familiar with the structure and contents of these documents, which will be key to the Reflexive Thematic Analysis (Chap. 3).

2.2.1 The Transhumanist Declaration[51]

The original *Transhumanism Declaration* was written in 1998 by an international group of authors including Doug Baily, Anders Sandberg, Gustavo Alves, Max More, Holger Wagner, Natasha Vita-More, Eugene Leitl, Bernie Staring, David Pearce, Bill Fantegrossi, den Otter, Ralf Fletcher, Tom Morrow, Alexander Chislenko, Lee Daniel Crocker, Darren Reynolds, Keith Elis, Thom Quinn, Mikhail Sverdlov, Arjen Kamphuis, Shane Spaulding, and Nick Bostrom. Throughout the years, the original Declaration has been modified multiple times by several authors and organisations.

The version of the *Transhumanist Declaration* used in this research was adopted by the H+ Board in March 2009. Overall, the Declaration is synthetical by essence, allowing us to know the fundamental elements of Transhumanism.

The Transhumanist Declaration contains eight articles which explore the topics of science and technology, the enhancement of human conditions, the misuse of new technologies, research efforts, the reduction of existential risks, policy-making guided by moral vision, the well-being of all sentience, and the favouring of wide personal choice.

[47] HUMANITY+, «Transhumanism», *Humanity+*, in https://www.humanityplus.org/transhumanism [11-6-2022].
[48] Cf. HUMANITY+, «The Transhumanist Declaration», 2009, *Humanity+*, in https://www.humanityplus.org/the-transhumanist-declaration [11-6-2022].
[49] Cf. HUMANITY+, «Transhumanist FAQ 3.0», 2001, *Humanity+*, in https://www.humanityplus.org/transhumanist-faq [2-12-2020].
[50] Cf. N. VITA- MORE, «The Transhumanist Manifesto (v.4)», 2020, *Humanity+*, in https://www.humanityplus.org/the-transhumanist-manifesto [11-6-2022].
[51] HUMANITY+, «The Transhumanist Declaration», 2009, *Humanity+*, in https://www.humanityplus.org/the-transhumanist-declaration [11-6-2022].

2.2.2 Transhumanist FAQ 3.0[52]

The FAQ was developed in the mid-1990s and published in 1998. It was established as a formal document through the inspirational work of transhumanists such as Alexander Chislenko, Max More, Anders Sandberg, Natasha Vita-More, Eliezer Yudkowsky, Arjen Kamphius, and many others. Over the years, this FAQ has been updated to provide an accurate description of Transhumanism. In 2001, the WTA (now H+) adopted the FAQ, and Nick Bostrom added important information concerning future scenarios. The present research uses *Transhumanist FAQ 3.0,* the version presented on the H+ website.

This text was the longest of the three transhumanist documents chosen for this work. *Transhumanist FAQ 3.0* is a text of major importance, as it discusses a wide range of transhumanist topics which help us to gain a greater understanding of *The Transhumanist Declaration.*

The structure and questions in *Transhumanist FAQ 3.0* are as follows:

- *Initial questions*:
  What is transhumanism? What is a posthuman? What is a transhuman? What are the reasons to expect all these changes? Won't these developments take thousands or millions of years? How can I use transhumanism in my own life? What if it doesn't work? How could I become a posthuman? Won't it be boring to live forever in a perfect world? How can I get involved and contribute?
- *Society and Politics*:
  Will new technologies only benefit the rich and powerful? Why transhumanists advocate human enhancement as ethical rather than pre-WWII eugenics? Aren't these future technologies very risky? Could they even cause our extinction? If these technologies are so dangerous, should they be banned? What can be done to reduce the risks? Shouldn't we concentrate on current problems, such as improving the situation of the poor, rather than putting our efforts into planning for the "far" future? What kind of society would posthumans live in? Will posthumans or superintelligent machines pose a threat to humans who aren't augmented? Is there any ethical standard by which transhumanists judge "improvement of the human condition"? Will extended life worsen overpopulation problems?
- *Technologies and Projections Biotechnology, Genetic engineering, Stem cells, and cloning:*
  What are they and what are they good for? What is molecular nanotechnology? What is superintelligence? What is virtual reality? What is cryonics? Isn't the probability of success too small? What is uploading? What is the singularity?
- *Transhumanism and Nature:*

[52] HUMANITY+, «Transhumanist FAQ 3.0», 2001, *Humanity+*, in https://www.humanityplus.org/transhumanist-faq [2-12-2020].

Why do transhumanists want to live longer? Isn't this tampering with nature? Will transhuman technologies make us inhuman? Isn't death part of the natural order of things? Are transhumanist technologies environmentally sound?

- *Transhumanism as a Philosophical and Cultural Viewpoint*:

  What are the philosophical and cultural antecedents of transhumanism? What currents are there within transhumanism? What kind of transhumanist art is there? How does transhumanism relate to religion? Won't things like uploading, cryonics, and AI fail because they can't preserve or create the soul?

2.2.3 The Transhumanist Manifesto v.4[53]

Written by Natasha Vita-More in 1993, *The Transhumanist Manifesto* is inspired by the 1983 *Transhuman Statement*. It was revised in 1998 (v.2), 2008 (v.3), and 2020 (v.4). The last version is the one used in this work.

Out of the three documents, *The Transhumanist Manifesto* is the only document authored by a single transhumanist, rather than a group. Therefore, it may be considered less significant than the other two documents. However, Humanity+ includes *The Transhumanist Manifesto* as one of the "original documents on Transhumanism".[54]

The Transhumanist Manifesto v.4 contains the following parts:

- Human Aging: The Option of Ageless Thinking
- A Vision for Shared Appreciation of Life and Purpose
- Life Extension & Expansion
- I Would Rather be a Transhuman than a Cyborg
- Transhumanist Rights, Freedom & Ecology
- Transhumanist Political Potential & Sensibility
- Beyond Prejudice/Toward Integrity
- The 1983 text of *Transhuman Statement (Manifesto)*

[53] N. VITA- MORE, «The Transhumanist Manifesto (v.4)», 2020, *Humanity+*, in https://www.humanityplus.org/the-transhumanist-manifesto [11-6-2022].

[54] HUMANITY+, «Transhumanism», *Humanity+*, in https://www.humanityplus.org/transhumanism [11-6-2022].

Common Themes Between the Governance of Human Genome Editing and Transhumanism

3

Abstract

To determine the common themes between Transhumanism and the global governance of human genome editing (HGE), the Reflexive Thematic Analysis (RTA) method was applied. This chapter examines the basics of RTA method and summarises the choices and outcomes of its application to selected texts on Transhumanism and selected texts on the global governance of HGE. In this research, Transhumanism is represented through the three documents which Humanity+ refers to as the original documents on Transhumanism. The texts selected to represent the global governance of HGE are the three documents published by WHO on the topic in 2021, the only framework for the global governance of HGE to date. After following the six steps of the RTA method, three common themes emerged. These commonalities lead to some implications for today's Bioethics.

Keywords

Reflexive thematic analysis • Transhumanism • Governance • Gene editing • World Health Organization • Health • Enhancement

3.1 Reflexive Thematic Analysis

Considering that Transhumanism aims to fundamentally improve the human condition "especially by developing and making widely available technologies",[1] along with the fact that transhumanists have been discussing the topic of gene editing for decades, perhaps Transhumanism is present in the global governance of

[1] HUMANITY+, «Transhumanism», Humanity+, in https://www.humanityplus.org/transhumanism [11-6-2022].

L. Santos, *Governance of Human Gene Editing and Transhumanism*,
Integrated Science 28, https://doi.org/10.1007/978-3-031-70863-3_3

HGE. Even if Transhumanism is not present as an organisation planning the global governance, transhumanist ideas are gradually gaining ground in academic, scientific, and cultural spheres, likely engaging the stakeholders and impacting the governance process.

Whilst constructing a plan for the global governance of emerging technologies such as HGE, Transhumanist ideas may appear. It might be the case that although the governance plan for HGE would not be aligned with all transhumanist proposals, it would still be likely to address some transhumanist-related topics. For example, a governance plan for a technique created half a century ago would most likely not consider enhancement-related issues. In contrast, today it would be unsuitable for the governance of gene editing to not address enhancement.

Considering the above, the leading questions of the present research are:

1. Are there common themes between Transhumanism and the plan for global governance of human genome editing? If yes, what are they?
2. Does a common topic mean a shared vision, or an agreement on the topic?
3. What are some of the implications for Bioethics that arise from the common themes between Transhumanism and the plan for global governance of HGE?

To answer the first and second questions, Reflexive Thematic Analysis (RTA)[2] was applied. The purpose of this qualitative research method is to develop *themes* across a dataset that address a research question. In RTA, "themes are actively constructed patterns (or meanings) derived from a data set that answer a research question, as opposed to mere summaries or categorizations of codes."[3] The theme is not merely a topic, as it implies a greater interpretation and integration of the

[2] Cf. V. BRAUN - V. CLARKE, «Using thematic analysis in psychology», Qualitative Research in Psychology 3/2 (2006), 77–101, in https://www.tandfonline.com/doi/abs/10.1191/1478088706qp063oa [31-8-2022]; Cf. V. BRAUN - V. CLARKE, «Thematic analysis», in APA handbook of research methods in psychology, Vol 2: Research designs: Quantitative, qualitative, neuropsychological, and biological, APA handbooks in psychology, American Psychological Association, Washington, D.C. 2012, 57–71; Cf. V. BRAUN - V. CLARKE, Successful Qualitative Research: A Practical Guide for Beginners, SAGE, Los Angeles 2013; Cf. V. BRAUN - V. CLARKE, «Reflecting on reflexive thematic analysis», Qualitative Research in Sport, Exercise and Health 11/4 (2019), 589–597, in https://doi.org/10.1080/2159676X.2019.1628806 [11-11-2022]; Cf. V. BRAUN - V. CLARKE, «One size fits all? What counts as quality practice in (reflexive) thematic analysis?», Qualitative Research in Psychology 18/3 (2021), 328–352, in https://doi.org/10.1080/14780887.2020.1769238 [11-11-2022]; Cf. V. BRAUN - V. CLARKE, Thematic Analysis: A Practical Guide, SAGE Publications Ltd, London and Thousand Oaks 2021; Cf. V. BRAUN - V. CLARKE, «Conceptual and design thinking for thematic analysis», Qualitative Psychology 9 (2022), 3–26.

[3] M. E. KIGER - L. VARPIO, «Thematic analysis of qualitative data: AMEE Guide No. 131», *Medical Teacher* 42/8 (2020), 846–854, in https://doi.org/10.1080/0142159X.2020.1755030 [31-8-2022], 1.

data. It captures an important aspect about the data in relation to the research question and portrays a level of patterned meaning within the data set.[4]

Presented below is the synthetic definition of the main concepts included in this method:

- *Data:* pieces of information selected for the analysis.
- *Code:* a succinct label that captures important features of the data that might be relevant to addressing the research question.
- *Subcode:* a label of minor importance that specifies a feature of the main code.
- *Inductive approach:* the researcher approaches the data without preconceived ideas in mind, allowing the data to determine the codes or themes.
- *Deductive approach:* the researcher approaches the data with specific questions in mind to code around, along with some pre-established ideas based on the theory or some existing knowledge.
- *Semantic approach:* a descriptive analysis of the data, aiming solely at identifying the explicit content or surface meanings of the data.
- *Latent approach:* an approach that goes beyond the descriptive level (what is written) attempting to identify hidden meanings, assumptions, or ideas that may shape the data.
- *Theme:* an actively constructed pattern or meaning derived from a data set which answers a given research question.
- *Theme's definition:* a clear formulation of the final theme, after a process of searching, reviewing, and refining the candidate themes.
- *Theme's name:* a short informative formulation of the final theme.
- *Theme's properties or elements:* sub-themes that further specify the content.
- *Report:* a selection of compelling extract examples and final analysis, relating the themes to the research question.

  The most widely accepted framework for conducting thematic analysis involves a six-step process, according to Braun and Clarke[5]:
- *Step 1—Familiarisation with the data:* reading and re-reading the data and noting down initial ideas.
- *Step 2—Coding:* generating succinct labels (codes) that capture important features of the data that might be relevant to addressing the research question. It involves systematically coding the entire dataset, with two or more rounds of coding. Next, the research collates all of the codes and all of the relevant data extracts for later stages of analysis.

---

[4] Cf. V. BRAUN - V. CLARKE, «Using thematic analysis in psychology», Qualitative Research in Psychology 3/2 (2006), 82.
[5] Cf. *Ibid.*, 87.

- *Step 3—Searching for themes:* grouping codes and collected data to create candidate (also called initial or potential) themes, gathering all data relevant to each potential theme. In this phase, the researcher develops broader patterns of meaning (candidate themes).
- *Step 4—Reviewing themes:* checking the candidate themes against the coded data and the entire dataset, in order to determine whether they tell a convincing story of the data while also addressing the research question. In this phase, themes are further developed, meaning that some candidate themes may be split, combined, or discarded.
- *Step 5—Defining and naming themes:* an ongoing analysis to refine each theme, along with the overall story that the analysis tells, generating clear definitions and names for each theme. The final theme should be the pattern of shared meaning underpinned by a central concept or idea. This phase involves developing a detailed analysis of each theme and deciding on an informative name for each theme.
- *Step 6—Producing the report:* the selection of compelling extract examples, the final analysis, relating back to the research question and literature, and the production of a scholarly report of the analysis.

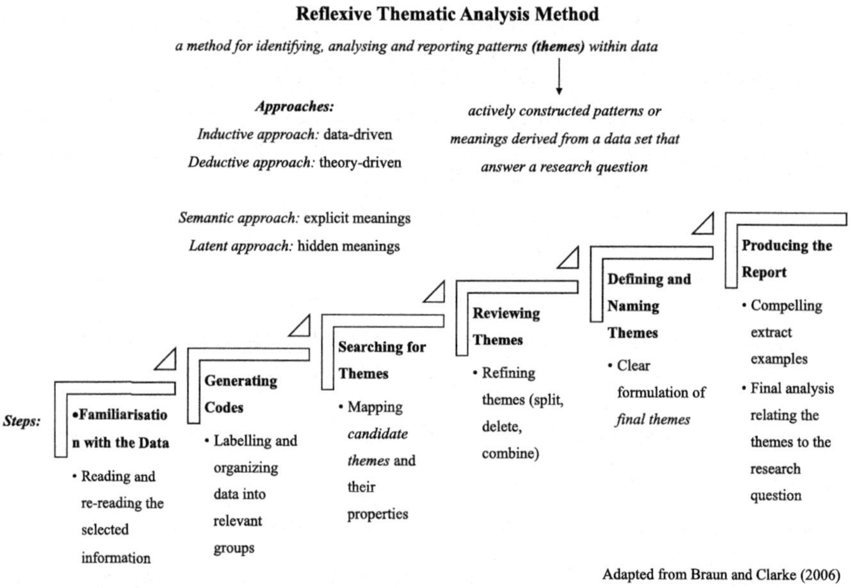

Fig. 3.1 Reflexive Thematic Analysis Method[6]

[6] L. SANTOS, *Common Themes Between Humanity+ Original Documents on Transhumanism and WHO Documents on the Global Governance of Human Genome Editing, Reflexive Thematic Analysis and Implications for Bioethics*, Ateneo Pontificio Regina Apostolorum, Rome 2023, 37.

Reflexive Thematic Analysis "is a time consuming process that evolves as the researcher navigates the different phases. This can lead to new interpretations of the data, which may in turn require further iterations of earlier phases."[7] This method implies going back and forth searching, analysing, relating, reorganising, and reviewing information until the final definition of themes is determined. Figure 3.1 synthetically shows the method and its six steps.

In the present work, data had to be selected in order to search for common themes between Transhumanism and the global governance of HGE. The selected texts to represent Transhumanism in this research are the three documents that Humanity+ refers to as the "original documents on transhumanism",[8] giving them primacy over other

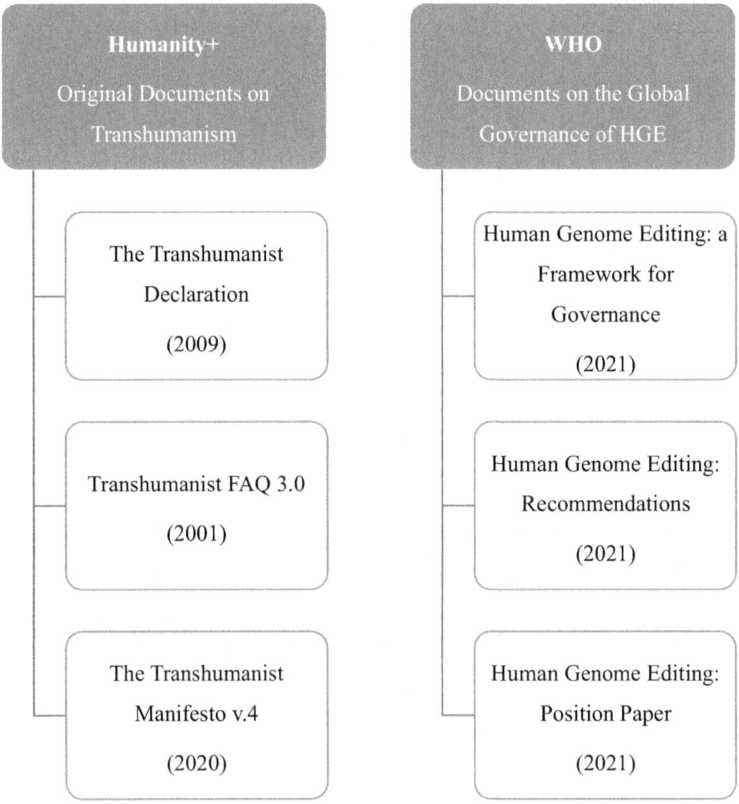

Fig. 3.2 Selected data

[7] D. BYRNE, «A worked example of Braun and Clarke's approach to reflexive thematic analysis», *Quality & Quantity* 56/3 (2022), 1391–1412, in https://doi.org/10.1007/s11135-021-01182-y [18-10-2022], 1398.

[8] HUMANITY+, «Transhumanism», *Humanity+*, in https://www.humanityplus.org/transhumanism [11-6-2022].

transhumanist works in order to represent the convergences among transhumanist thought. Until the conclusion of this research in 2023, these texts were:

- *The Transhumanist Declaration*[9]
- *Transhumanist FAQ 3.0*[10]
- *The Transhumanist Manifesto*[11]

The selected texts to represent the global governance of HGE are the three documents published by WHO on the topic in 2021. This choice considers the international influence of the WHO and, most importantly, that these texts are the only existing *global* plan for the governance of HGE to date.

- *Human genome editing: a framework for governance*[12]
- *Human genome editing: recommendations*[13]
- *Human genome editing: position paper*[14]

Therefore, six texts divided into two datasets (Humanity+texts and WHO texts) constitute the selected data in this research.

ATLAS.ti software was used to upload the 6 texts and label the quotations according to the 8 codes and 16 subcodes I chose to apply to the dataset. The 6 documents, containing about 260 pages, resulted in 755 coded quotations. Afterwards, ATLAS.ti provided an Excel with the quotations classified by codes. I

[9] Cf. HUMANITY+, «The Transhumanist Declaration», 2009, *Humanity+*, in https://www.humanityplus.org/the-transhumanist-declaration [11-6-2022].

[10] Cf. HUMANITY+, «Transhumanist FAQ 3.0», 2001, *Humanity+*, in https://www.humanityplus.org/transhumanist-faq [2-12-2020].

[11] Cf. N. VITA- MORE, «The Transhumanist Manifesto (v.4)», 2020, *Humanity+*, in https://www.humanityplus.org/the-transhumanist-manifesto [11-6-2022].

[12] Cf. WHO EXPERT ADVISORY COMMITTEE ON DEVELOPING GLOBAL STANDARDS FOR GOVERNANCE AND OVERSIGHT OF HUMAN GENOME EDITING, *Human genome editing: a framework for governance* WHO, Geneva 2021, in https://www.who.int/publications/i/item/9789240030060 [13-7-2021].

[13] Cf. WHO EXPERT ADVISORY COMMITTEE ON DEVELOPING GLOBAL STANDARDS FOR GOVERNANCE AND OVERSIGHT OF HUMAN GENOME EDITING, *Human genome editing: recommendations*, WHO, Geneva 2021, in https://www.who.int/publications-detail-redirect/9789240030381 [13-7-2021].

[14] Cf. WHO EXPERT ADVISORY COMMITTEE ON DEVELOPING GLOBAL STANDARDS FOR GOVERNANCE AND OVERSIGHT OF HUMAN GENOME EDITING, *Human genome editing: position paper*, WHO, Geneva 2021, in https://www.who.int/publications-detail-redirect/9789240030404 [13-7-2021].

reread all the quotations to further analyse the content of each one. Then, I searched for themes, starting with 8 candidate themes. Lastly, I worked on their supporting quotations and elements, following the verification questions[15] and guidelines by Braun and Clarke, until I reached my 3 final themes with their elements.

Below is a summary of the main information concerning my application of the RTA method in this research.[16] Each box corresponds to one of the six steps of the RTA. The first row in each box presents the *Step* of the method. The second row presents a brief *Description* of its implications. The third row defines the *Choices* made within the boundaries of the corresponding RTA step. The fourth row illustrates the *Outcomes* of the step (Tables 3.1, 3.2, 3.3, 3.4, 3.5 and 3.6).

Table 3.1 My application of the RTA method—Step 1

1. Familiarisation with the data	
Description	• Selecting data according to the research objective • Reading and re-reading the information
Choices	• H+Original Documents on Transhumanism • WHO Documents on Global Governance of HGE
Outcomes	• **6 documents (263 pages) divided into 2 sets:** **3 H+documents** *(Declaration, FAQ, Manifesto)* **3 WHO documents** *(Framework, Recommendations, Position Paper)*

[15] Cf. V. BRAUN - V. CLARKE, «Thematic analysis», in APA handbook of research methods in psychology, Vol 2: Research designs: Quantitative, qualitative, neuropsychological, and biological, APA handbooks in psychology, American Psychological Association, Washington, D.C. 2012, 57–71, 65.

[16] The full Reflexive Thematic Analysis (RTA) can be consulted in the published doctoral dissertation excerpt L. Santos, *Common Themes Between Humanity+ Original Documents on Transhumanism and WHO Documents on the Global Governance of Human Genome Editing: Reflexive Thematic Analysis and Implications for Bioethics*, Ateneo Pontificio Regina Apostolorum, Rome 2023, 84.

Table 3.2 My application of the RTA method - Step 2

2. Generating codes	
Description	• Labelling and organising data in relevant groups Generation of codes and subcodes Application of the codes to the select data
Choices	• Hybrid approach *Which ideas are present in H+ documents?* Inductive and Semantic approaches to create the codes and apply them to H+ documents *Are these ideas present in WHO documents?* Deductive and Latent approaches to apply the created codes to WHO documents
Outcomes	• **755 coded quotations according to these 8 main codes and 16 subcodes:** **1. Science and Technology** 1a. tech benefits 1b. direct evolution/redesign nature 1c. gene editing **2. Enhanced human condition** **3. Risks of misuse of tech** **4. Research efforts, decisions, and implementation** 4a. social order that decides/implements (governance) 4b. risk–benefit approach 4c. tech risks 4d. public debate **5. Urgent priorities to be funded** 5a. reduction of existential risks 5b. preservation of life and health 5c. alleviation of suffering 5d. funding **6. Policy making guided by moral vision** 6a. policies 6b. individual rights 6c. solidarity, inclusion and no eugenics 6d. equality, social justice 6e. responsibility future generations/sustainability **7. Well-being of all sentience** **8. Wide personal choice**

Table 3.3 My application of the RTA method—Step 3

3. Searching for themes	
Description	• Mapping candidate themes and their elements
Choices	• Identification of common elements between the 2 sets of texts • Presentation of the most relevant coded quotations from H+ Documents and WHO Documents regarding the topic of each of the 8 main codes • 1 candidate theme for each main code
Outcomes	• **8 candidate themes and their elements:** 1. **The significant impact of technology** Increasing availability Impact of HGE 2. **The scenario of human enhancement** Addition of desired new traits Concerns: equality, liberty, social acceptance 3. **The unscrupulous and destructive use of technology** Recognition of this risk Call for solutions 4. **Governance to minimise risks and maximise benefits** Will to reduce risks and maximise the benefits of new technologies Need for good global governance Public debate 5. **The allocation of funds according to priorities** Funding as a governance tool Preservation of life and health Reduction of suffering 6. **Policy making guided by values** Autonomy and individual rights Equality and solidarity Responsibilities towards future generations 7. **Beyond personal well-being** Well-being as a goal Extension to more beings 8. **Wide personal choices in health and reproduction** Respect for the wishes of individuals Protection of people who cannot express their wishes

Table 3.4 My application of the RTA method—Step 4

4. Reviewing themes	
Description	• Refining candidate themes (splitting, deleting, combining themes and elements) • Using the verification questions
Choices	• Grouping candidate themes • Analysis of candidate themes' elements, and then deleting, combining, and splitting candidate themes and their elements accordingly
Outcomes	• **4 groups of candidate themes** **Technology** (candidate themes 1 and 3) **Well-being and Enhancement** (candidate themes 2 and 7) **Governance** (candidate themes 4, 5, 6) **Liberty** (candidate theme 8)

Table 3.5 My application of the RTA method—Step 5

5. Defining and naming themes	
Description	• Giving a clear formulation (definition) of final themes • Giving a short title (name) to each final theme
Choices	• Line: What (HGE), What for (intended uses), How to manage (GG)
Outcomes	**3 final themes (definitions,** *names,* and elements) **1. The significant impact of new technologies such as HGE** *WHAT: Impact of HGE Technology* Increasing Power and Availability Potential Benefits Risks: Technical and Misuse **2. HGE for health, well-being, and enhancement** *WHAT FOR: Health, Well-being, and Enhancement* Priorities: Life, health, well-being Possibility: Enhancement **3. Global Governance to minimise risks and maximise benefits of HGE** *HOW TO MANAGE: Governance to maximise benefits* Elements: Research, Funds, Public debate, Policies Values: Individual rights/autonomy in health and reproduction. Protection of people who cannot express themselves. Equality-Solidarity

Table 3.6 My application of the RTA method—Step 6

6. Producing the report	
Description	• Compelling extract examples supporting the final themes • Mental map recommended • Final analysis relating the themes to the research question
Choices	• Explanation of steps taken accompanied by tables, figures, and mental maps
Outcomes	• **Research questions 1 and 2 answered**

The first research question was: *Are there common topics between Transhumanism and the plan for global governance of HGE? If yes, what are they?* Three common themes resulted from this application of the RTA method: *1. The significant impact of new technologies such as HGE, 2. HGE for health, well-being and enhancement, 3. Global Governance to minimise risks and maximise benefits of HGE.* The three final themes with their definition, names and properties or elements are displayed in Fig. 3.3.

The second research question was: *Does a common topic mean a shared vision, or an agreement on the topic?* More than mere topics, the RTA method results in common themes. A theme in RTA is a pattern or meaning derived from data. To make this concept clearer: global governance, for example, is simply a topic, while the RTA theme related to global governance presents the meaning and elements that appeared as a pattern related to this topic in the selected data (see definition and elements of theme 3). Therefore, RTA themes are far more valuable for knowledge than mere topics.

It should be noted that identifying a common theme does not imply agreement on every detail. For example, both WHO and H+sets of texts address enhancement, meaning an intervention that improves the average functioning of a human capacity. However, that does not mean that the texts agree on every aspect of enhancement. For instance, H+ considers enhancement a priority, while the WHO does not. Furthermore, the expectations and concerns that H+and the WHO have regarding enhancement are not identical: one perspective emphasises liberty, whilst the other emphasises equality. However, the elements of RTA final theme 2 present a shared pattern regarding enhancement.

To give another example, theme one refers to the impact of HGE technology, taking into account its increasing power and availability, its potential benefits, and

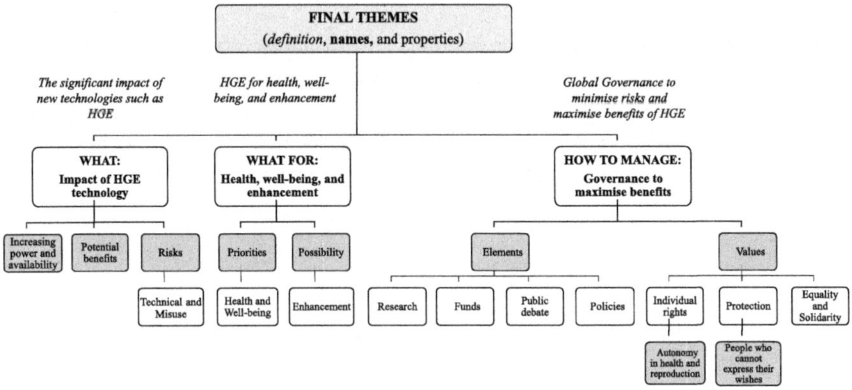

Fig. 3.3 Final themes[17]

[17] L. SANTOS, *Common Themes Between Humanity+ Original Documents on Transhumanism and WHO Documents on the Global Governance of Human Genome Editing. Reflexive Thematic Analysis and Implications for Bioethics*, Ateneo Pontificio Regina Apostolorum, Rome 2023, 230.

its associated risks. However, transhumanists maintain higher expectations concerning new technologies compared to the expectations of the WHO. Throughout H+ texts, technology is perceived as a means to direct evolution. On the other hand, WHO texts focus primarily on the use of science and technology to promote the health of humanity. The WHO is more concerned about ensuring effective governance of new technologies, therefore its texts pay greater attention to the risks and challenges of new technologies than the texts of H+.

In short, a common *topic* does not constitute a shared vision or agreement. Instead, a common *theme* derived from RTA constitutes a shared pattern of meaning, even if this meaning does not imply an identical view or consensus regarding every detail related to the theme.

Presented below is a more extensive explanation of each final theme.

3.1.1 Theme 1: The Significant Impact of New Technologies Such as HGE

This first theme is composed of three elements: first, the increasing power and availability of HGE technologies. Second, the potential benefits of HGE technologies. And third, the risks, both the technical risks and the risks of misuse associated with HGE technologies. The impact of HGE technologies is reviewed from both positive and negative perspectives.

According to the selected H+ texts, "Humanity stands to be profoundly affected by science and technology in the future."[18] Transhumanists are confident that "technology has a huge role to play"[19] in the future of humanity. Technology is continually growing in its potential and availability, and "it is not an especially bold conjecture that the next 50 years will see at least as much change and that the state of technology in the mid-twenty-first century will be quite wondrous by present standards."[20]

Common technologies such as computers and airplanes, which some believed initially to be technically impossible or socially irrelevant, have become a significant part of our daily lives.[21] In *Transhumanist FAQ 3.0*, the authors affirm that "It would require an abrupt reversal of current trends, an unexpected deceleration, in order for the changes that many transhumanists foresee not to happen within the twenty-first

18 HUMANITY+, «The Transhumanist Declaration» 1, 2009, Humanity+, in https://www.humanityplus.org/the-transhumanist-declaration [11-6-2022].

19 HUMANITY+, «Transhumanist FAQ 3.0», 2001, Humanity+, in https://www.humanityplus.org/transhumanist-faq [2-12-2020].

20 *Ibid.*

21 Cf. *Ibid.*

century".[22] According to H+ texts, technology is increasing in its potential and availability. Provided that civilisations continue to develop and prosper, and unless forcibly prevented, many will soon choose to explore transhumanist options.[23]

Regarding the WHO texts, these were created precisely because the WHO is aware of the increasing potential and availability of gene editing technologies. The plan for the global governance of HGE acknowledges the rapidly evolving nature of emerging technologies. WHO states that "an ongoing process is needed that takes into account relevant technical, societal, economic, legal and ethical developments and adapts governance measures accordingly."[24]

Both H+ texts and WHO texts display feelings of excitement towards the advancements that will be made possible by gene editing. H+ texts state: "Many transhumanist aims can be pursued with present technologies".[25] Regarding HGE technology, many transformational opportunities are listed in WHO texts as potential benefits: "new strategies for diagnosis, treatment and prevention of genetic disorders; new avenues to treat infertility; new ways to promote disease resistance; contribution to vaccine development and enhanced knowledge of human biology."[26] Furthermore, WHO texts mention that "somatic human genome editing has already been undertaken (…) to address HIV, sickle-cell disease and transthyretin amyloidosis,"[27] and germline editing "contributes to deepen our understanding of the role of specific genes and processes in early human development, physiology and diseases."[28]

Both the opportunities and risks associated with HGE should be taken seriously.[29] Regarding the latter, WHO texts address both technical risks and the risks of misusing HGE for unscrupulous purposes. It is important that research efforts carefully consider how to reduce the risks and maximise the benefits of HGE.[30] The WHO notes that "there are important areas of ongoing uncertainty as to potential benefits and risks, and gaps in scientific understanding in such key domains as off-target effects and long-term risks."[31] Some risks to be considered include:

---

[22] *Ibid.*

[23] Cf. *Ibid.*

[24] WHO Expert Advisory Committee on Developing Global Standards for Governance and Oversight of Human Genome Editing, *Human genome editing: recommendations*, WHO, Geneva 2021, in https://www.who.int/publications-detail-redirect/9789240030381 [13-7-2021], 6.

[25] Humanity+, «Transhumanist FAQ 3.0», 2001, Humanity+, in https://www.humanityplus.org/transhumanist-faq [2-12-2020].

[26] WHO Expert Advisory Committee on Developing Global Standards for Governance and Oversight of Human Genome Editing, *Human genome editing: a framework for governance*, WHO, Geneva 2021, in https://www.who.int/publications/i/item/9789240030060 [13-7-2021], v.

[27] *Ibid.*

[28] *Ibid.*

[29] Cf. Humanity+, «The Transhumanist Declaration», 2009, Humanity+, in https://www.humanityplus.org/the-transhumanist-declaration [11-6-2022], 6.

[30] Cf. *Ibid.* 4.

[31] WHO Expert Advisory Committee on Developing Global Standards for Governance and Oversight of Human Genome Editing, *Human genome editing: a*

> The harms could include incorrect on-target events such as unwanted insertions or deletions, chromosome damage as well as off-target events. When genome editing is carried out on a stem cell line from which single cells are expanded to give a clonally derived stem cell line, it is possible to test for such events. However, if the genome editing is being done on many millions of cells simultaneously, it will be very challenging to show that all are free of such potentially harmful events. In addition, unwanted genetic alterations, such as chromosome rearrangements, which can lead to cell overproliferation can be tolerated by somatic cells and lead to tumours.[32]

It is worth noting that the WHO does not address the technical risks of HGE as a reason to suggest that it should be abandoned. Rather, the WHO emphasises that applications require proceeding with caution, more research and trials, and further improvements. The goal is to reduce risks and maximise benefits: "The balance between benefit and harm, safety and speed, and innovation and access is relevant to all of human genome editing."[33]

The dangers of the misuse of technology are recognised throughout both H+ and WHO texts. The WHO affirms that "(...) developments at the frontier of new scientific disciplines such as genomics, epigenetics, gene editing, artificial intelligence, and big data, all of which pose transformational opportunities but also risks to global health."[34] On their behalf, transhumanist authors state: "We recognize that humanity faces serious risks, especially from the misuse of new technologies. There are possible realistic scenarios that lead to the loss of most, or even all, of what we hold valuable."[35] Both H+ and WHO texts show agreement with the general affirmation made in *The Transhumanist Declaration*: "Although all progress is change, not all change is progress."[36]

Emerging technologies are introducing into our world great possibilities along with enormous challenges that could lead to the extinction of our species. According to transhumanists, "The gravest existential risks facing us in the coming decades will be of our own making".[37] Furthermore, "The coming technological transitions may be the most important challenge that humanity will ever face. The entire future of intelligent life on Earth may depend on how we handle it."[38] The WHO Committee members similarly mention that they "saw and heard evidence of challenges associated with rogue clinics, medical travel, as well as the

---

framework for governance, WHO, Geneva 2021, in https://www.who.int/publications/i/item/9789240030060 [13-7-2021], v.

[32] *Ibid.*, 18.

[33] *Ibid.*, ix.

[34] WHO in *Ibid.*, v.

[35] HUMANITY+, «The Transhumanist Declaration», 2009, Humanity+, in https://www.humanityplus.org/the-transhumanist-declaration [11-6-2022], 3.

[36] *Ibid.*

[37] HUMANITY+, «Transhumanist FAQ 3.0», 2001, Humanity+, in https://www.humanityplus.org/transhumanist-faq [2-12-2020].

[38] *Ibid.*

reporting of illegal, unregistered, unethical or unsafe research and other activities including the offer of unproven so-called therapeutic interventions."[39]

The paragraph above portrays examples of the quotations considered in building theme one. Upon analysing all related quotations and completing the RTA process, the final definition of theme 1 is *The significant impact of new technologies such as human genome editing*. This theme constitutes the "WHAT" of the overall narrative: the vision of technology (especially HGE) and its impact, opportunities, and risks. Within the RTA process, after the definition and verification of the theme have been completed, the theme needs to be given a concise name. Therefore, the chosen name for theme 1 is *The impact of HGE technology.*

3.1.2 Theme 2: HGE for Health, Well-Being, and Enhancement

This theme is comprised of the uses of HGE for health, well-being, and enhancement purposes. Within the selected texts, health and well-being are connected, and both are considered as priorities. Regarding enhancement, this is regarded as a priority by transhumanists, but not by the WHO Committee. However, the WHO texts address the topic of enhancement on numerous occasions. Overall, the convergence seen within the H+and WHO texts is that enhancement is a possibility that lies on the horizon.

As a result, the title of this theme expresses that HGE's intended uses include health, well-being, and enhancement. The theme's properties are divided into priorities and possibilities, in order to show that the texts agree on health and well-being as immediate priority, and enhancement as a future possibility.

The concepts of health and well-being are included in both H+and WHO texts and are often connected. H+texts states: "Transhumanism is the world's strongest advocacy for a positive future in changing the human condition of disease, poverty and prejudice toward health, well-being, and prosperity."[40] The WHO framework for governance affirms that "The technology of human genome editing can be used to expand human knowledge, improve human health and contribute to both collective well-being and the common good."[41]

It is important to note that health and well-being are also linked to the alleviation of suffering. According to H+, "When transhumanists seek to extend human

[39] WHO Expert Advisory Committee on Developing Global Standards for Governance and Oversight of Human Genome Editing, *Human genome editing: a framework for governance*, WHO, Geneva 2021, in https://www.who.int/publications/i/item/9789240030060 [13-7-2021], 2.

[40] N. Vita- More, «The Transhumanist Manifesto (v.4)», 2020, Humanity+, in https://www.humanityplus.org/the-transhumanist-manifesto [11-6-2022].

[41] WHO Expert Advisory Committee on Developing Global Standards for Governance and Oversight of Human Genome Editing, *Human genome editing: a framework for governance*, WHO, Geneva 2021, in https://www.who.int/publications/i/item/9789240030060 [13-7-2021], 12.

life, they are not trying to add a couple of extra years at a care home spent drooling at one's shoes. The goal is more healthy, happy, productive years."[42] Transhumanists also make clear that "When transhumanists say they want to extend lifespans, what they mean is that they want to extend healthspans."[43]

Regarding the meaning of the word enhancement, Humanity+ texts portray enhancement as the wished improvement of the human condition through technological means. "Transhumanism is a philosophical and cultural movement concerned with promoting responsible ways of using technology to enhance human capacities and to increase the scope of human flourishing."[44] According to transhumanists, technology can and should be used to improve ourselves: "We are not limited to traditional humanistic methods, such as education and cultural development. We can also use technological means (…)[45] and "Just as we use rational means to improve the human condition and the external world, we can also use such means to improve ourselves, the human organism."[46]

As for the WHO, WHO texts on the global governance of HGE address enhancement through various dimensions. The WHO acknowledges the unanswered questions regarding the interpretation of health: "How will distinctions be drawn between disease and disability prevention, and therapy?",[47] and enhancement: "Will this be the introduction of traits that are beyond ordinary human capacity and health? Or, will this be the introduction of traits beyond those usually present in the human species?"[48] WHO texts describe enhancement in different ways, such as the addition of desirable new traits, the improvement of present traits, and interventions not intended for health nor that increase human performance. For example: "At some time in the future, somatic human genome editing might be able to be used to *improve typical and perfectly adequate traits* or to *add new traits*."[49] Furthermore, when addressing heritable HGE, the WHO says it "could be used to reduce the number of people born with so-called undesirable traits or increase the number born with *improved* and possibly *new* so-called *desirable traits*."[50] Lastly, the WHO recognises that both somatic and heritable gene editing may be implemented for enhancement purposes.

Another commonality that exists between H+ and WHO texts regarding enhancement is that neither of the two completely opposes it. Instead, both identify

[42] HUMANITY+, «Transhumanist FAQ 3.0», 2001, Humanity+, in https://www.humanityplus.org/transhumanist-faq [2-12-2020].

[43] *Ibid.*

[44] *Ibid.*

[45] *Ibid.*

[46] *Ibid.*

[47] WHO EXPERT ADVISORY COMMITTEE ON DEVELOPING GLOBAL STANDARDS FOR GOVERNANCE AND OVERSIGHT OF HUMAN GENOME EDITING, *Human genome editing: a framework for governance*, WHO, Geneva 2021, in https://www.who.int/publications/i/item/9789240030060 [13-7-2021], 27.

[48] *Ibid.*

[49] *Ibid.*, 26. The italic is mine.

[50] *Ibid.* The italic is mine.

concerns regarding its risks, while urging readers to reflect on some historically and socially accepted enhancements. For instance, one of the WHO texts states: "attention to the full range of temporary and permanent medical and biological enhancements already in use, ranging from nutritional supplements to strength-enhancing steroids to cosmetic surgeries".[51]

To conclude this section, theme 2 constitutes the "WHAT FOR" of the overall story, by reviewing the priorities and possibilities associated with the use of HGE. Considering the complete RTA process, the definition of theme 2 is *Human genome editing for health, well-being, and enhancement*. Furthermore, the concise name for theme 2 is *Health, well-being, and enhancement.*

3.1.3 Theme 3: Global Governance to Minimise Risks and Maximise Benefits of HGE

This final theme is centred on the global governance of HGE. The most important elements of governance found in the selected WHO and H+ texts include *research*, *funds*, *public debate*, and *policies*. The values that inspire governance measures commonly addressed by H+ and WHO can be grouped into three concepts: *autonomy*, *protection*, and *equality*.

Governance is perceived as a means to minimise the risks and maximise the benefits of HGE. The WHO states that it is "important to consider the overall, or net, potential benefit and risk profile when determining how and why human genome editing might be used."[52] Furthermore, the WHO quotes that "The balance between benefit and harm, safety and speed, and innovation and access is relevant to all of human genome editing."[53] With regards to Transhumanism, Transhumanists believe that a "world order characterized by peace, international cooperation, and respect for human rights would much improve the odds that the potentially dangerous applications of some future technologies can be controlled".[54] Moreover, regarding governance and its elements, *The Transhumanist Declaration* states:

> *Research* effort needs to be invested into understanding these prospects. We need to carefully *deliberate* how best to reduce risks and expedite beneficial applications. We also need *forums* where people can constructively discuss what should be done, and a *social order* where responsible *decisions* can be *implemented*.[55]

---

[51] *Ibid.*, 26.
[52] *Ibid.*, 48.
[53] *Ibid.*, ix.
[54] HUMANITY+, «Transhumanist FAQ 3.0», 2001, Humanity+, in https://www.humanityplus.org/transhumanist-faq [2-12-2020].
[55] HUMANITY+, «The Transhumanist Declaration», 2009, Humanity+, in https://www.humanityplus.org/the-transhumanist-declaration [11-6-2022], 4. The italic is mine.

WHO texts affirm that "Good governance is an *iterative, ongoing process* that includes *mechanisms* for regular revision in light of technical, practical and ethical *developments* and changes in *societal views and values*."[56] The WHO acknowledges that the global governance of new technologies involves multiple stakeholders and "will best be achieved by taking advantage of the full range of individuals and organizations able to influence or control the direction of research and possible future uses."[57] The entirety of Part 4 of the WHO framework is dedicated to outlining the tools, institutions, and processes associated with the governance of HGE. These components are listed as follows:

- Declarations, treaties, conventions, legislation and regulations, judicial rulings, and ministerial decrees
- Conditions on research funding
- Moratoria
- Accreditation, registration or licensing
- National science and medicine societies and institutions
- Patents and licences/Professional self-regulation
- Public advocacy and activism
- Research ethics guidelines and research ethics review
- Collaboration with publishers and conference organizers
- Education and training of researchers and clinicians[58]

WHO says that "Governance also includes forces to shape the direction and conditions of research and practice."[59] Research efforts are shaped not only by formal mechanisms such as legislation, regulations, or judicial opinion, but also by informal mechanisms such as ethical, social, and professional norms. For example, various factors such as industrial best practices, peer reviews, ethics approvals, funding, and insurance coverage strongly influence research development.

With regards to funding, both transhumanists and members of the WHO Committee also recognise this as something that strongly influences research, decisions and implementation. The following quotations exemplify how H+ texts perceive funding as a tool for governance: "Reduction of existential risks, and development of means for the preservation of life and health, the alleviation of grave suffering, and the improvement of human foresight and wisdom should be pursued as urgent priorities, and heavily funded."[60] Specifically, "Gene therapy, stem cell research,

[56] WHO Expert Advisory Committee on Developing Global Standards for Governance and Oversight of Human Genome Editing, *Human genome editing: a framework for governance*, WHO, Geneva 2021, in https://www.who.int/publications/i/item/9789240030060 [13-7-2021], 10. The italic is mine.

[57] *Ibid.*, 28.

[58] Cf. *Ibid.*, 28–39.

[59] *Ibid.*, 10.

[60] Humanity+, «The Transhumanist Declaration», 2009, Humanity+, in https://www.humanityplus.org/the-transhumanist-declaration [11-6-2022], 5.

therapeutic cloning, and other areas of medicine that have the potential to deliver these benefits deserve a high priority in the allocation of research monies."[61] Furthermore, transhumanist texts state that "the rate of progress depends on levels of funding, which in turn depends on hard-to-predict economic and political factors."[62]

In its turn, WHO states "one powerful source of governance arises from decisions determining priorities and setting funding rules."[63] Funding is an especially crucial factor during the initial development process of technology. For instance, WHO notes that "Some of the first somatic human genome editing protocols that have been licensed cost more than US$ 500 000 per patient."[64] Furthermore, funding is a necessary aspect for maintaining governance mechanisms. This idea is portrayed in the statement: "Substantial resources will be required for sustained international collaboration."[65]

As the selected texts point out, actions such as research, international cooperation, public engagement, and other governance-related mechanisms require funding to take place. Since resources are limited, the allocation of funding obeys an explicit or implicit list of priorities. Therefore, the allocation of funds can be defined as a governance tool.

Regarding public debate, the following quotations illustrate its place in governance. *The Transhumanist Declaration* states: "We also need forums where people can constructively discuss what should be done".[66] Additionally, Transhumanism wants to be included in public debate: "In addition to scholarly studies of the threats and their possible countermeasures, public awareness must be raised to enable a more informed debate of our long-term options."[67] Furthermore, the *Transhumanist FAQ* affirms that "Awareness of transhumanist ideas is spreading. Transhumanism is undergoing the transition from being the preoccupation of a fringe group of intellectual pioneers to becoming a mainstream approach (…) [It] is no longer an insight confined to a few handfuls of techno-savvy visionaries."[68]

---

[61] HUMANITY+, «Transhumanist FAQ 3.0», 2001, Humanity+, in https://www.humanityplus.org/transhumanist-faq [2-12-2020].
[62] *Ibid.*
[63] WHO EXPERT ADVISORY COMMITTEE ON DEVELOPING GLOBAL STANDARDS FOR GOVERNANCE AND OVERSIGHT OF HUMAN GENOME EDITING, *Human genome editing: a framework for governance*, WHO, Geneva 2021, in https://www.who.int/publications/i/item/9789240030060 [13-7-2021], 32.
[64] *Ibid.*, 18.
[65] WHO EXPERT ADVISORY COMMITTEE ON DEVELOPING GLOBAL STANDARDS FOR GOVERNANCE AND OVERSIGHT OF HUMAN GENOME EDITING, *Human genome editing: recommendations*, WHO, Geneva 2021, in https://www.who.int/publications-detail-redirect/9789240030381 [13-7-2021], 7.
[66] Cf. HUMANITY+, «The Transhumanist Declaration», 2009, Humanity+, in https://www.humanityplus.org/the-transhumanist-declaration [11-6-2022], 4.
[67] HUMANITY+, «Transhumanist FAQ 3.0», 2001, Humanity+, in https://www.humanityplus.org/transhumanist-faq [2-12-2020].
[68] *Ibid.*

Within WHO texts, the principle of *responsible stewardship of science* implies "a commitment to align the processes and outcomes of human genome editing research with the values, needs and expectations of society, as identified through participatory approaches involving various publics."[69] Throughout the texts, attention is given to the need for public debate, its challenges and how to improve participation. Increased public engagement "would help shape the research and policy agenda so that these reflect public values, perspectives, experiences, interests and priorities."[70]

In terms of policy-making guided by a moral vision, the following quotation from *The Transhumanist Declaration* exemplifies its importance for governance:

> Policy making ought to be guided by responsible and inclusive moral vision, taking seriously both opportunities and risks, respecting autonomy and individual rights, and showing solidarity with and concern for the interests and dignity of all people around the globe. We must also consider our moral responsibilities towards generations that will exist in the future.[71]

WHO texts stress that "Good governance is value-based and principle-driven. It promotes public trust by ensuring public values and viewpoints are carefully considered as part of the policy-making process."[72] The Committee proposes that the following values and principles be considered to inform *how* decisions are made[73]:

- Openness, transparency, honesty and accountability
- Responsible regulatory stewardship
- Responsible stewardship of science
- Responsible stewardship of research resources.

And that the following values and principles be considered to inform *what* decisions are made[74]:

[69] WHO Expert Advisory Committee on Developing Global Standards for Governance and Oversight of Human Genome Editing, *Human genome editing: a framework for governance*, WHO, Geneva 2021, in https://www.who.int/publications/i/item/9789240030060 [13-7-2021], 13.

[70] WHO Expert Advisory Committee on Developing Global Standards for Governance and Oversight of Human Genome Editing, *Human genome editing: recommendations*, WHO, Geneva 2021, in https://www.who.int/publications-detail-redirect/9789240030381 [13-7-2021], 16.

[71] Humanity+, «The Transhumanist Declaration», 2009, Humanity+, in https://www.humanityplus.org/the-transhumanist-declaration [11-6-2022], 6.

[72] WHO Expert Advisory Committee on Developing Global Standards for Governance and Oversight of Human Genome Editing, *Human genome editing: a framework for governance*, WHO, Geneva 2021, in https://www.who.int/publications/i/item/9789240030060 [13-7-2021], 11.

[73] Cf. *Ibid.*, 13–14.

[74] Cf. *Ibid.*

- Inclusiveness
- Caution
- Fairness
- Social justice
- Non-discrimination
- Equal moral worth
- Respect for persons
- Solidarity
- Global health justice

The WHO believes that "Careful attention to these values and principles is imperative to create trust in the choice of governance mechanisms and policy options."[75] However, it is worth noting that while proposing these two sets of values and principles, the Committee also reports the need to search for values: "To address the absence of a set of WHO principles that can be consistently used in WHO guidance documents, the Committee recommends developing a set of WHO ethical values and principles that can be consistently applied across all WHO activities, including human genome editing."[76]

Upon analysing the selected texts and quotations that discuss values, along with following the steps of the RTA method, I concluded that the principal values can be divided into three distinct groups. The first value is respect for individual rights and autonomy. The second value is the protection of people who cannot express their wishes, especially those of future generations. The third value is equality-solidarity.

Concerning the first group of common values (individual rights and autonomy), both sets of texts call to respect persons and their rights, emphasising their autonomy in health and reproduction. Liberty is understood as the accomplishment of the wishes of the individual. Presented below are some quotations concerning individual rights and autonomy, beginning with *The Transhumanist Declaration*:

> We favour allowing individuals **wide personal choice** over how they enable their lives. This includes use of techniques that may be developed to assist memory, concentration, and mental energy; life extension therapies; reproductive choice technologies; cryonics procedures; and many other possible human modification and enhancement technologies.[77]

According to transhumanists, individuals should have the choice to live or to die, to enhance or not to enhance, and have access to the necessary means to achieve

[75] *Ibid.*, x.

[76] WHO Expert Advisory Committee on Developing Global Standards for Governance and Oversight of Human Genome Editing, *Human genome editing: recommendations*, WHO, Geneva 2021, in https://www.who.int/publications-detail-redirect/9789240030381 [13-7-2021], 19.

[77] Humanity+, «The Transhumanist Declaration», 2009, Humanity+, in https://www.humanityplus.org/the-transhumanist-declaration [11-6-2022], 8. The bold is mine.

their personal desires. From a transhumanist perspective, "people ought to have the right to determine how they want to live, for how long, and if they want to opt in or out of biomedical anti-ageing therapies."[78] "Saving lives (of those who want to live) is ethically important",[79] and "Ideally, everybody should have the right to choose when and how to die – or not to die."[80] Overall, Transhumanism places great emphasis on individual freedom of choice and autonomy.

This approach to individual freedom of choice and autonomy is not endorsed in the same way by WHO texts. However, the Transhumanist and WHO texts are close to each other when referring to the autonomy of the individual regarding their health and reproductive choices. The WHO Committee, following the principle of *respect for persons*, implies "a commitment to respect the wishes of competent individuals regarding the most intimate aspects of their lives, including their health and their reproductive options."[81]

The second group of common values (the protection of people who cannot express their wishes), is intimately connected to the first group of values (individual rights and autonomy). People who cannot express their wishes need protection. Here, the value of autonomy appears in a different form. When discussing people who cannot express their wishes, the WHO texts refer largely to future generations. Transhumanists defend procreative liberty, but stress that it should not lead to any risk of child neglect or child abuse. WHO texts state that the principle of *respect for persons* implies respecting people's wishes in health and reproduction, along with a "commitment to promote the best interests of individuals who are not competent to make decisions for themselves."[82]

The third group of common values is equality-solidarity. Among the WHO values and principles to be considered that inform *what* decisions are made, one value is closely related to autonomy and individual rights (*Respect for persons*), another value is closely related to protection and safety (*Caution*), but most of the values are related to equality and solidarity (*Inclusiveness, Fairness, Social justice, Non-discrimination, Equal moral worth, Solidarity* and *Global health justice*).

H+ also expresses a deep concern for inequality. Transhumanist authors declare that the improvement they dream of seeing within humanity should not resemble pre-World War II eugenics, since Transhumanism calls for enhancements and highly free and tolerant societies. "In addition to condemning the coercion

[78] N. VITA- MORE, «The Transhumanist Manifesto (v.4)», 2020, Humanity+, in https://www.humanityplus.org/the-transhumanist-manifesto [11-6-2022].
[79] HUMANITY+, «Transhumanist FAQ 3.0», 2001, Humanity+, in https://www.humanityplus.org/transhumanist-faq [2-12-2020].
[80] *Ibid.*
[81] WHO EXPERT ADVISORY COMMITTEE ON DEVELOPING GLOBAL STANDARDS FOR GOVERNANCE AND OVERSIGHT OF HUMAN GENOME EDITING, *Human genome editing: a framework for governance*, WHO, Geneva 2021, in https://www.who.int/publications/i/item/9789240030060 [13-7-2021], 14.
[82] *Ibid.*

involved in such policies, transhumanists strongly reject the racialist and classist assumptions on which they were based (…)"[83] According to transhumanist authors, enhancements should be available to everyone, with no discrimination to access and no discrimination as an outcome.

To conclude this section, this third final theme provides the "HOW" for the overall story. After considering the "WHAT" (the impact of HGE technologies), along with the "WHAT FOR" (the intended uses of HGE), this third common theme refers to "HOW TO MANAGE" human gene editing, concerning the global governance of HGE. The definition of theme 3 is *Global Governance to minimise risks and maximise benefits of HGE.* Theme 3 comprises key elements of governance (research, funds, public debate, and policies) and the predominant values in the governance of HGE. These values are divided into three distinct groups. Firstly, the respect for individual rights (largely meaning autonomy in health and reproduction). Secondly, the protection of people who cannot express their wishes (especially those of future generations). Thirdly, equality-solidarity. Lastly, the concise name tor Theme 3 is *Governance to maximise benefits.*

3.2 Implications for Bioethics

The identification of common themes has provided answers for the first two research questions. *1. Are there common themes between Transhumanism and the plan for global governance of human genome editing? If yes, what are they? 2. Does a common topic mean a shared vision, or an agreement on the topic?*

Once common themes have been identified, one can focus on the current implications for Bioethics and what can be done. Before the present situation-common themes between Transhumanism and the global governance of HGE-what should be done? What does all this mean for Bioethics? What are the tasks for bioethicists? Therefore, the third and last research question is: *What are some of the implications for Bioethics that arise from the common themes between Transhumanism and the plan for global governance of HGE?*

In this work, any theme's implication for Bioethics is comprised of two aspects. Firstly, the *recognition of the current situation* for Bioethics following the identification of common themes between Transhumanism and the GG of HGE. Secondly, the *identification of the tasks* for Bioethics that arise from the current situation. Beyond listing or analysing the situation emerging from the common themes, this book tries to identify and work on the tasks considering the present scenario. For each common theme presented below, one example of an implication for Bioethics (*situation* and *task*) is given.

---

[83] Humanity+, «Transhumanist FAQ 3.0», 2001, Humanity+, in https://www.humanityplus.org/transhumanist-faq [2-12-2020].

Table 3.7 Common themes and implications for bioethics

Common themes	Implications for bioethics
The impact of HGE technology • Increasing Power and Availability • Potential Benefits • Risks: Technical and Misuse	**Situation:** ***Technology in the past, present, and future of Bioethics*** **Task:** ***To address*** • The increasing power of HGE and the human condition • The increasing availability of technology and DIY mentality • Technology at the service of desires
HGE for health, well-being, and enhancement • Priorities: Health-Well-being • Possibility: Enhancement	**Situation:** ***The Limitations of "Therapy (allowed) vs Enhancement (forbidden)"*** **Task:** ***To work on a better bioethical framework*** • Object, intentions, and circumstances of enhancement • Proposals by non-transhumanists • Principle of Proportionality
GG to maximise benefits of HGE • Elements: Research, Funds, Public debate, Policies • Values: – Individual rights-autonomy (in health and reproduction) – Protection (of people who cannot express themselves) – Equality-Solidarity	**Situation:** ***Bioethics' role in the GG of HGE*** **Task:** ***To refine the predominant values in the GG of HGE*** • Autonomy- reconnecting liberty and attention to coercion • Protection- to all the vulnerable • Equality- Intrinsic

Table 3.7 presents a synthesis of the three common themes (with their definitions and elements) along with examples of the implications that they introduce for Bioethics.

Part 2 will further develop these implications for Bioethics.

The task for Bioethics related to Theme 1 refers to the need to address the increasing power of HGE and the human condition, the increasing availability of technology and DIY mentality, and the issue of technology at the service of desires. This is covered in Chap. 4, titled *Addressing Possibilities, Desires and Questions.*

The implication for Bioethics related to Theme 2 implies recognising the limitations of the paradigm "therapy (allowed) versus enhancement (forbidden)" and offering a better bioethical framework. This is done in Chaps. 5, 6 and 7.

The implication for Bioethics related to Theme 3 comprises, on the one hand, the role of Bioethics in the global governance of HGE, addressed in Chap. 8. On the other hand, there is the task of refining the predominant values of the global governance of HGE. This is the object of Chap. 9. Lastly, Chap. 10 specifies some bioethical notions regarding the different types of HGE. This is part of Bioethics' contribution to the global governance of HGE.

Part II

Tasks for Today's Bioethics

An abridged version was published as a paper: SANTOS, L., "Transhumanism and Global Governance of Human Genome Editing: Common Themes and Implications for Bioethics", *Medicina y Ética* 34/4 (2023), 1089–1163, in https://doi.org/10.36105/mye.2023v34n4.05.

4 Addressing Possibilities, Desires, and Questions

Abstract

A common theme that resulted from the Reflexive Thematic Analysis (RTA) method applied to selected WHO and Humanity+ texts is *the impact of human genome editing technology*. The birth of Bioethics is closely linked to technological progress. Contemporary Bioethics has been tasked with answering various emerging questions in order to serve as a bridge to the future. Humans have been co-evolving with technology since prehistoric times. However, thanks to human genome editing (HGE), humans have moved beyond external interventions and toward a transformation from the inside out. Considering the current situation, bioethicists should address the possibilities, desires, and questions that emerge alongside the innovation of new technologies such as HGE. A task for contemporary Bioethics includes expanding its range of considerations to include the following: firstly, the increasing potential of HGE and how it may affect the human condition. Secondly, the increasing availability of technology combined with the do-it-yourself (DIY) mentality. And lastly, the trend of technologies at the service of personal desire.

Keywords

Transhumanism • Gene editing • Human nature • DIY • Biohacking • Technology

L. Santos, *Governance of Human Gene Editing and Transhumanism*,
Integrated Science 28, https://doi.org/10.1007/978-3-031-70863-3_4

4.1 The Increasing Power of HGE and the Human Condition

A common theme that resulted from the RTA method applied to selected WHO and Humanity+ texts is the *impact of human genome editing technology*.[1] The field of life sciences is becoming more a matter of "editing" life rather than "reading" it, designing creation rather than decoding its content. One peculiarity of our time is the fact that technology deeply influences humanity in significant ways. This concept is a core transhumanist idea, portrayed in the very first article of *The Transhumanist Declaration*: "Humanity stands to be profoundly affected by science and technology".[2] According to leading transhumanist Nick Bostrom, the combination of technologies—virtual reality, preimplantation of genetic diagnosis, genetic engineering, pharmaceuticals that improve memory, concentration, wakefulness, and mood, performance-enhancing drugs, cosmetic surgery, sex change operations, prosthetics, anti-ageing medicine, and closer human-computer interfaces—could all profoundly transform the human condition.[3]

Gene editing is increasingly perceived as a tool that has the potential to profoundly affect humanity, rather than being simply a new medical technique. Therefore, the growing interest in the general public towards CRISPR is becoming more and more noticeable. Transhumanists tend to accompany technological advances with singular expectations, and as a result, consider CRISPR to be one of the most powerful technological inventions in recent years. Transhumanists believe that effecting heritable alterations is key to generating successful transhuman change, highlighting the desirability and inevitability of germline and enhancing gene therapies.[4] Therefore, current Bioethics is tasked with refocusing the question of the human condition, while taking into consideration the increasing potential of HGE and the spread of transhumanist ideas.

Can certain applications of HGE change the human condition and affect human identity? Will this depend on whether HGE has therapeutic versus enhancement *intention*? Will the difference depend on the addition of any *novelty to the human genetic pool*? What about xenotransplantation, a concept that tends to be morally accepted? Will the answer depend on the *amount* of human or non-human genetic changes? Could the key factor be the *kind* of genetic alteration? Or will the key

---

[1] The full Reflexive Thematic Analysis (RTA) can be consulted in the published doctoral dissertation excerpt L. SANTOS, *Common Themes Between Humanity+ Original Documents on Transhumanism and WHO Documents on the Global Governance of Human Genome Editing*, Ateneo Pontificio Regina Apostolorum, Rome 2023.

[2] HUMANITY+, «The Transhumanist Declaration», 2009, Humanity+, in https://www.humanityplus.org/the-transhumanist-declaration [11-6-2022], 1.

[3] Cf. N. BOSTROM, «A History of Transhumanist Thought», in M. RECTENWALD - L. CARL (edd.), *Academic Writing Across the Disciplines,* Pearson Longman, New York 2011, 12.

[4] R. RANISCH, «When CRISPR Meets Fantasy: Transhumanism and the Military in the Age of Gene Editing», in W. HOFKIRCHNER - H.-J. KREOWSKI (edd.), *Transhumanism: The Proper Guide to a Posthuman Condition or a Dangerous Idea?*, Cognitive Technologies, Springer International Publishing, Cham 2021, 111–120, 113.

factor depend on the overall *effects of HGE*? Where will the line be drawn between a modified (and maybe enhanced) human and a chimaera? What makes a human being "human"? This last is an old philosophical question. However, today it is becoming a scientific question as well. What makes a being belong to the human species in the gene editing era?

The concept of species is not free of controversy within theoretical biology, and yet, species taxonomy plays a fundamental role in the organisation of biological entities. The 'species' category generally includes three factors: *genealogy*, the *similarity of the genome*, and the *potential for interbreeding*.[5] With regards to the first factor, "the evolutionary lineage of anatomically modern humans is framed in taxonomic ranks that are deployed in an ancestry-based timeline. *H. sapiens* are the only extant species of the genus *Homo* (which commenced around 2.5 million years ago), belonging to the order of Primates (85–55 Ma), to the class of Mammalia (220 Ma) and to the kingdom Animalia (600 Ma)."[6] As for the second factor (the similarity of the genome), humans share a complete set of genetic material, and this is a fundamental unity that exists amongst all members of our species. In terms of the third factor (the potential for interbreeding), the notion of species also includes members of a population that can be recognised as potential mates for reproductive purposes.

Many would agree that the human being is not defined only by genes. And yet, it is an undeniable fact that humans are embodied beings. But, embodied with what kind of body? The *human* body. Does this matter? Considering all the possibilities that genetic engineering has to offer, what makes our human body belong to this species? And will this answer remain the same through time?

Transhumanism challenges our understanding of humanity, but at the same time, it introduces opportunities to deepen and clarify the old explanations of human identity. For instance, anthropologies rooted in classical Aristotelian-Thomist philosophical tradition emphasise man as being a rational animal. However, what if a type of non-human body is also capable of rationality? Or, what if a hybrid possesses rationality? "What does it mean to say that the human being is a rational animal? If rationality remains the only specific difference that identifies our kind, then would a human-chimp hybrid who is rational be no different in essence from one of us?"[7] What if it becomes technically possible to customise the body at a genomic level? Will some people demand a right to that? Do we truly have the right to a kind of biological or "morphological freedom",[8] as claimed by transhumanists? In such a scenario, how will belonging to the human species be defined?

---

[5] Cf. J. RUEDA, «Genetic enhancement, human extinction, and the best interests of posthumanity», *Bioethics* (2022), 1–10, in https://onlinelibrary.wiley.com/doi/abs/10.1111/bioe.13085 [13-2-2023], 2.

[6] *Ibid.*

[7] N. P. G. AUSTRIACO, *Biomedicine and Beatitude: An Introduction to Catholic Bioethics*, The Catholic University of America Press, Washington, D.C. 2021^2, 402.

[8] Cf. A. SANDBERG, «Morphological Freedom», in *The Transhumanist Reader*, John Wiley & Sons, Ltd, Chichester 2013, in https://doi.org/10.1002/9781118555927.ch5 [14-4-2020], 56–64.

"Where does biology become metaphysics? Where do we cross the threshold between actualizing potentials we have always had and becoming new kinds of beings?"[9] These questions are no longer fantastical. With the technical possibilities of interference in the natural processes of evolution of whole species, answers are more urgent than ever before.

The former statement is especially true if Bioethics is called to illuminate and propose moral paths, rather than simply to evaluate and propose policies after a given technology has been developed. In other words, an abstract reflection on the ethics of technology is not enough, nor is the mere listing of concerns regarding rapidly evolving technologies. If Bioethics limits itself to proposing general principles for the approval or disapproval of a specific technology, it will not be fulfilling its function. Altogether, Bioethics exists in order to explore difficult and fundamental questions. The European Group on Ethics in Science and New Technologies stated that the ethics of the future needs to be "an ambitious, proactive, daring and public ethics, an ethics that asks the hard, painful or tedious questions."[10]

The field of Bioethics has constructed principles accepted by the majority of countries around the world. These principles are portrayed in the *Universal Declaration of Bioethics and Human Rights,*[11] a declaration approved by all 193 UNESCO member states. This Declaration was created without a globally shared definition of human nature or the human condition. Although it was possible to achieve agreement in 2005, will it be possible to come to a consensus regarding how far we can go in human genome editing without sharing a common notion of what it means to be human? How can we discuss to what extent HGE might affect humanity without a shared notion of the human condition?

In 1997, UNESCO adopted the *Universal Declaration on the Human Genome and Human Rights* (UDHGHR).[12] During the 1990s, feasible applications of HGE were far more limited than they are today. The first article of the UDHGHR states that in a symbolic sense, the human genome is "the heritage of humanity".[13] Back then, this could have been interpreted as something intangible. However, in the present, interpretations of the human genome are far more complex. Is there a human genetic patrimony that is intangible, or is the human genome a collection

[9] B. P. GREEN, «Transhumanism and Catholic Natural Law: Changing Human Nature and Changing Moral Norms», in *Religion and Transhumanism: The Unknown Future of Human Enhancement*, ABC-CLIO, Santa Barbara 2014, 201–215, 207–208.

[10] EUROPEAN GROUP ON ETHICS IN SCIENCE AND NEW TECHNOLOGIES, *Values for the Future*, Publications Office of the European Union, Brussels 2021, in http://op.europa.eu/en/web/eu-law-and-publications/publication-detail/-/publication/6d9879f7-8c55-11eb-b85c-01aa75ed71a1 [16-11-2021], 16.

[11] Cf. UNESCO, «Universal Declaration on Bioethics and Human Rights», Declaration pg.80, in http://portal.unesco.org/en/ev.php-URL_ID=31058&URL_DO=DO_TOPIC&URL_SECTION=201.html [8-6-2019].

[12] Cf. UNESCO, «Universal Declaration on the Human Genome and Human Rights», 1997, in http://portal.unesco.org/en/ev.php-URL_ID=13177&URL_DO=DO_TOPIC&URL_SECTION=201.html [3-12-2019].

[13] *Ibid.* 1.

of data that can be modified? In the 90s the human genome was regarded as a common heritage, yet today it seems to exist as a dataset that can be read, manipulated, and even hacked.[14] Even if the expression of "the heritage of humanity"[15] is maintained, this change in the way of looking at life has consequences for what can be done with the genetic heritage. For example, the interpretation of heritage may shift from something that is untouchable, to something that should be protected. Is it better to leave this heritage unchanged? Can genetic inheritances be protected, while at the same time being modified or improved in certain aspects? Article 1 of the UDHGHR also states that "The human genome underlies the fundamental unity of all members of the human family, as well as the recognition of their inherent dignity and diversity."[16] The Declaration states that the human genome underlines and helps us to recognise humanity. However, what if the reference genome is not as fixed as we considered it to be in the 1990s?

The question of how HGE might affect the human condition will probably not arise as a direct and literal question, but rather take the form of questions such as: Is it time to change the reference human genome? Is it more appropriate to leave the idea of the standard genome and move toward some kind of incremental reference genome? Should we stay with what genome we were given or should we improve upon it? If so, to what extent? This idea is mentioned in the following quotation by Ballouz et al.: "As population datasets become broader and individual datasets become deeper, it appears to be time to think about both the virtues of the current reference and our potential options to replace or augment it."[17] The authors question whether is it time to change the reference genome and conclude that a "switch to a consensus genome would not be a transformational change to current practice and would provide a far from perfect standard, but because it would offer incremental, broad-based, and progressive improvement, we believe that it is time to make this change."[18]

To conclude this section regarding HGE and the human condition, the main questions identified in this work are grouped as follows:

- *Global issue*: Can some people decide to modify the genetic heritage of humanity? Does heritage imply something that cannot be touched? Is it possible to agree on the applications of human gene editing without consensus on what it means to be human? Is it possible to reach a significant global agreement on the definition of human identity?
- *Possible changes in the human condition*: Could the human condition be changed one day? Could some applications of HGE reach the point of altering human

[14] Cf. L. SANTOS, «Biohacking and Transhumanism: what and why», *Relectiones* 9 (2022), 40–53, in https://doi.org/10.32466/eufv-rel.2022.9.743.40-53 [11-11-2022], 40.
[15] UNESCO, «Universal Declaration on the Human Genome and Human Rights», 1997, in http://portal.unesco.org/en/ev.php-URL_ID=13177&URL_DO=DO_TOPIC&URL_SECTION=201.html [3-12-2019], 1.
[16] *Ibid.*
[17] S. BALLOUZ - A. DOBIN - J. A. GILLIS, «Is it time to change the reference genome?», *Genome Biology* 20/1 (2019), 159, in https://doi.org/10.1186/s13059-019-1774-4 [13-2-2020], 8.
[18] *Ibid.*

identity? If yes, which factors will be the most relevant—the intention of the gene editing (therapeutic or enhancement), amount of genes changed, types of gene editing, amount of original human genes remaining, or the outward effects of the gene change?

- *Average, incremental or customised reference genome*: Should we change what is currently considered the reference human genome? Should we leave behind the idea of the basic or average genome and move towards an incremental reference genome? What if it becomes technically possible to customise the body at a genomic level? What makes the human body human? Does the type of physical body matter? In an era of gene editing, what makes us belong to the same species?
- *Definition of the human condition in the current age*: What is the human condition, what is human nature, and what constitutes human identity? How will belonging to the human species be defined in the age of gene editing and transhumanist aspirations? Is there a better way to explain what makes us human, one that is up to the challenge presented by the current era of gene editing?

The ability to answer all the above questions is beyond the scope of this work, however, the identification of these sets of questions about the human condition is a step forward in research and a priority for contemporary Bioethics. Bioethical reflections should refocus the question of the human condition considering the increasing potential of HGE and the spread of transhumanist ideas.

Whether or not the human condition is susceptible to change at this essential level, I agree with Austriaco that "twentieth-first century bioethics is going to struggle primarily with questions regarding human identity."[19] It might even be the case that humanity as a species will not change no matter how many genetic modifications are made. However, the combination of Transhumanism aspirations along with the rise of more advanced techniques for HGE leaves the question of the human condition as a challenging task for current Bioethics.

As technologies advance and introduce new possibilities, they introduce new questions as well. We should allow ourselves to be questioned. There is no point in rushing to offer insufficient answers. Transhumanism and HGE greatly challenge our understanding of technology, biology, and humanity. Transhumanism and HGE portray the difficulty associated with agreeing on these topics, and the need to deepen our understanding of them. This challenge is not necessarily a bad thing, and in fact, it can serve as a good opportunity for us to explore different perspectives. Transhumanism and gene editing invite us to find new ways to explain things that we used to take for granted. Even true definitions can be improved. This does not mean that these definitions are wrong. Instead, it is important to acknowledge that new ideas, technologies, and social trends have the potential to shed light upon new aspects, urging us to dig deeper and find better ways to explain the same reality.

It is important to listen and attempt to understand the desires and fears of our time. The issues related to HGE are not due to the knowledge and control we

[19] N. P. G. Austriaco, *Biomedicine and Beatitude: An Introduction to Catholic Bioethics*, The Catholic University of America Press, Washington, D.C. 2021 (2), 402.

have over our genes (which can be extremely beneficial). Rather, these issues are the results of the desires we have and/or the means we employ to achieve them. And yet, to assess whether or not these desires and means are truly beneficial for humanity (in other words, ethical), it is necessary to define the human condition—which is precisely what is being questioned in new ways by HGE and by Transhumanism. Therefore, a priority task for today's Bioethics is to identify and work towards answering the emerging questions concerning the human condition.

4.2 The Increasing Availability of Technology and the DIY Mentality

Significant attention should be given to the increasing availability of HGE combined with the popular *do-it-yourself* (DIY) mentality. Historically, Bioethics has directed its reflections almost exclusively towards health researchers and medical staff. Nowadays, not only is the potential of gene editing increasing but so is its types of users. With a growing number of biohackers and DIY communities, we are witnessing how "ordinary people" are using the large amount of available information on the internet to conduct their own HGE experiments. The layperson can obtain the necessary HGE tools for an accessible price, conduct personal experiments, and then share this process on social media with an aim of inserting high-tech science into everyday life.[20] As biohacking[21] and *do-it-yourself biology* (DIYbio) gain popularity as a social movement, we cannot ensure that emerging technologies will be exclusively in the hands of researchers and medical professionals. This concept is commonly referred to as the emergence of *open science* and *the democratisation of science*.[22]

An example that portrays this mentality is the case of Keoni Gandall, an individual who performed his first cloning experiment at the age of 12. When Gandall was in 7th Grade, his biology teacher noticed his strong interest in the subject

[20] Cf. T. J. KARLOVITZ, «The Democratization of Technology – and Its Limitation», in T. DIRSEHAN (ed.), *Managing Customer Experiences in an Omnichannel World: Melody of Online and Offline Environments in the Customer Journey*, Emerald Publishing Limited 2020, 13–25, in https://doi.org/10.1108/978-1-80043-388-520201004 [27-11-2020], 13.

[21] DIY science is a varied movement. To understand better the diversity of this movement, Cf. F. FERRETTI, «Mapping do-it-yourself science», *Life Sciences, Society and Policy* 15/1 (2019), 1, in https://doi.org/10.1186/s40504-018-0090-1 [14-3-2023]; A. DELFANTI, *Biohackers. The Politics of Open Science*, Pluto Press, London 2013; K. GRUBER, «Biohackers», *EMBO reports* 20/6 (2019), e48397, in https://www.embopress.org/doi/full/10.15252/embr.201948397 [27-11-2020].

More specifically on DIY and genetics: Cf. H. LEDFORD, «Biohackers gear up for genome editing», *Nature News* 524/7566 (2015), 398, in http://www.nature.com/news/biohackers-gear-up-for-genome-editing-1.18236 [27-11-2020]; Cf. P. J. ZETTLER - C. J. GUERRINI - J. S. SHERKOW, «Regulating genetic biohacking», *Science* 365/6448 (2019), 34–36, in https://www.science.org/doi/10.1126/science.aax3248 [14-3-2023]; Cf. C. J. GUERRINI ET AL., «Core values of genomic citizen science: results from a qualitative interview study», *BioSocieties* 17/2 (2022), 203–228.

[22] Cf. A. GIANNNKOULOPOULUS - L. LIMNIATI - D. HONORATO, *Portrayal of Biohackers in news and technological websites*, in (2017), 292–298, 292.

and allowed him to order a home bacterial transformation kit.[23] Further down the road, and without completing formal training or an academic degree, Gandall began working at Stanford University and led the *Free Genes Project*,[24] a group that sought to make DNA biotechnology more accessible to the general public.

There are examples of biohackers who have conducted self-experimentation and modified their own bodies. Gabriel Licina, for instance, managed to achieve night vision by injecting Chlorin e6 into his eyes. Furthermore, Neil Harbisson has been coined as the first government recognised cyborg, after implanting an antenna in his skull to help him to hear colours. Harbisson was born with achromatopsia, a condition characterised by the absence of colour vision. Lastly, Tim Cannon implanted a chip the size of a deck of cards in his arm, allowing him to transfer information such as body temperature onto his personal Android device.[25]

Within the field of augmentation technology, an individual by the name of Rich Lee stays informed about government projects, university projects, and DIY biohacking projects. In 2018, Rich Lee was interviewed by Newton Lee, Chairman of the California Transhumanist Party. During the interview, Rich Lee discussed biohacking as a solution to the poor affordability of high technologies: "(…) transhuman technology will be a product for the rich. If you are not rich but want these advantages for yourself or your children, you may have only one alternative: become a biohacker."[26] Rich Lee also expresses biohackers' opposition to governance over the body: "I am not willing to wait 10 or 20 years for some government to approve a certain intervention related to my transcendence. I do not even observe their right to approve or disapprove the things I wish to do with my body. I don't care about economic impact evaluations, ethics review boards, or social impact studies."[27] In Lee's view, when technologies become available, "they will be used no matter what the popular or legal opinion is. If not by your country, then by another. If not on the open markets then in the underground markets."[28]

Perhaps the most famous biohacker is Josiah Zayner, who previously worked for NASA. In October 2017, Zayner became the first person to modify his own genes using CRISPR live on Facebook. The particular modification that Zayner pursued aimed to increase muscular strength. According to him, this goal was scientific, but also cultural. Zayner's logic follows the idea that if people can modify their bodies with tattoos, piercings, and operations to obtain a desired appearance, their self-determination should also be respected in the case of genetic alterations. Furthermore, Zayner made available to the world the necessary tools to follow

[23] Cf. K. GRUBER, «Biohackers», EMBO reports 20/6 (2019), in https://www.embopress.org/doi/full/10.15252/embr.201948397 [27-11-2020], e48397, 1–2.

[24] Cf. «The FreeGenes Project», *stanford.freegenes.org*, in https://stanford.freegenes.org/ [9-2-2023].

[25] Cf. A. GIANNNKOULOPOULUS - L. LIMNIATI - D. HONORATO, *Portrayal of Biohackers in news and technological websites*, 294.

[26] R. LEE, «Biohacktivism, Cyborgasm, and Transhumanism», in N. LEE (ed.), *The Transhumanism Handbook*, Springer International Publishing, Cham 2019, 499–501, 500.

[27] *Ibid.*

[28] *Ibid.*

his steps. He published an online *Guideline to human CRISPR* and began selling 20-dollar kits including DNA that supports muscular growth.[29]

Biohackers attempt to democratise science, and many share a common motivation: they feel that advanced biotechnologies such as CRISPR should be available for people to do what they want rather than remain restricted to a monopoly of academics and pharmaceutics.[30] However, the idea that the individual should solve problems on their own may also increase the phenomenon of using technology for individual solutions, instead of taking into account social responsibilities. "This is what happens, for instance, when oocyte cryopreservation is proposed to allow a woman a rapid professional career and possibly a pregnancy at an older age instead of developing a system of daycare and assistance for working mothers."[31]

Considering the examples presented above regarding biohackers and DIY communities, we can see how ordinary people are using the extensive amount of information available online and obtaining the necessary tools to attempt gene editing at an accessible price. These individuals perform experiments and then share the corresponding information on social media, aiming to make science and high-tech methodologies a part of everyday life.[32] The point is that if our view towards life and technology changes, our actions will change accordingly. For instance, if life is viewed as a book or a machine, the following statement can be made: "If something is encoded, you can crack the code. (…) If you can crack the code, you can hack the code!"[33] A concept that long belonged within the field of computer science (hacking), is now related to the field of life (bio) sciences. Without great difficulty, life is being conceived as a set of data, DNA as a code, the body as a machine, and the mind as software… all of which are susceptible to being read, manipulated, and now even hacked.[34]

Transhumanists advocate for enhancement technologies to be widely accessible. "The transhumanist agenda, which is to make such enhancement options safely *available to all persons*, will become increasingly relevant and practical in the coming years as these and other anticipated technologies come online."[35] H+ director

[29] Cf. P. Benanti, *Digital age. Teoria del cambio d'epoca. Persona, famiglia e società*, San Paolo Edizioni, Milano 2020, 104–106.
[30] Cf. *Ibid.*
[31] Cf. International Bioethics Committee (IBC), «Report of the IBC on updating its reflection on the Human Genome and Human Rights», Paris 2 October 2015, in https://unesdoc.unesco.org/ark:/48223/pf0000233258 [14-2-2020], 11.
[32] Cf. T. J. Karlovitz, «The Democratization of Technology – and Its Limitation», in T. Dirsehan (ed.), *Managing Customer Experiences in an Omnichannel World: Melody of Online and Offline Environments in the Customer Journey*, Emerald Publishing Limited 2020, 13–25, in https://doi.org/10.1108/978-1-80043-388-520201004 [27-11-2020], 13.
[33] Dr. Joon Yun in S. Jasanoff, *Can Science Make Sense of Life?*, New Human Frontiers, Polity Press, Cambridge 2019, 172.
[34] Cf. L. Santos, «Biohacking and Transhumanism: what and why», *Relectiones* 9 (2022), in https://doi.org/10.32466/eufv-rel.2022.9.743.40-53 [11-11-2022], 40–53.
[35] N. Bostrom, «A History of Transhumanist Thought», in M. Rectenwald - L. Carl (edd.), Academic Writing Across the Disciplines, Pearson Longman, New York 2011, 12. The italic is mine.

Natasha Vita-More affirms that the do-it-yourself model strongly exemplifies transhumanist behaviour.[36] Furthermore, Brickley explains, "(…) biohackers are active participants who not only embrace and intellectually contribute to the philosophies of transhumanist thought, but actively engage with the ideas espoused in transhumanism in a physical way."[37]

Bioethics should not direct its reflections only towards researchers and medical staff, as doing so would ignore the reality that CRISPR is an easily accessible technology to the general public. The increasing availability of HGE technologies made accessible to a wide variety of users combined with the DIY mentality is a growing issue that contemporary Bioethics must address.

4.3 Technology at the Service of Desires

In order to propose effective measures for ethical living, bioethicists must understand the mindset behind new realities made possible *by* new technologies. Going back even further, bioethicists must also understand the hopes and desires that made possible the creation of a specific technology. As Jasanoff explains: "Through technology, human societies articulate their hopes, dreams, and desires while also making material instruments for accomplishing them."[38] For instance, Benanti writes about the emergence of the cyborg: "The birth of the cyborg is inextricably linked to both cyborg thinking and the existence of technologies capable of shaping and implementing these ideas."[39] The same case could be made for CRISPR babies, biohackers, and enhanced or modified beings. In *The Transhumanism Handbook,* it is written: "We have the potential, we have the ideas, we have the skills, and now we also have the technology."[40] Desire and ideas precede technology, or at least precede the acceptance and spread of certain technologies. Conscious regard should be given to the combination of technologies and the mentality behind them.

When evaluating our view towards science and technology, it is necessary to understand the changes that are taking place in modern society. What do the people of today expect from technology? What are their fears? What are their dreams? Does the transhumanist vision influence our expectations? WHO documents on the

[36] Cf. N. VITA- MORE, «History of Transhumanism», in N. LEE (ed.), *The Transhumanism Handbook*, Springer International Publishing, Cham 2019, 49–61, 34.

[37] L. BRICKLEY, «Bodies without Borders: The Sinews and Circuitry of "folklore+"», *Western Folklore* 78/1 (2019), 5–38, in https://www.jstor.org/stable/26864140 [10-7-2021], 12.

[38] S. JASANOFF, *The Ethics of Invention: Technology and the Human Future*, W. W. Norton & Co Inc, New York 2016, 242–243.

[39] P. BENANTI, *The Cyborg: corpo e corporeità nell'epoca del post-umano*, Studi e ricerche. Sez. teologica, Cittadella, Città di Castello 2012, 143. The translation is mine. The original follows: 'la nascita del cyborg è indissolubilmente legata tanto ad un pensiero cyborg che all'esistenza di tecnologie in grado di formare e realizzare queste idee.'.

[40] J. A. MOSSBRIDGE, «Leave Nobody Behind: The Future of Societies», in A. STOEL (ed.), *The Transhumanism Handbook*, Springer International Publishing, Cham 2019, 613–617, 167.

global governance of HGE have considered issues often addressed by Transhumanism, such as enhancement, germline gene editing, and biohacking. This does not mean that WHO agrees with Transhumanism on these topics. Instead, it is important to understand that the WHO framework for global governance of HGE would not be effective if it ignored the transhumanist trend. Similarly, Bioethics today is expected to acknowledge transhumanist desires.

Independent from the specific desires behind technology, two aspects of desires warrant significant attention. The first one is that today, the individual's desires are often not called into question. The social mindset that prevails throughout individualist societies seems to endorse this logic: If you desire to have a child, technology should give you one. What do prospective parents wish for? Echoing the Nuffield Council on Bioethics, some people want a child who is genetically related to them, healthy, and including or not including specific characteristics.[41] And it seems technology should fulfil these wishes. If someone desires a body with a particular image for themselves, technology can be used to help them achieve that too. If someone wants to increase their chances of excelling in sports, academics, work or interrelationship goals, technology is expected to be at hand for that as well.

A second aspect of desire related to technology is that the present time seems to be characterised by the desire for "more"—for more life, more health, more power, more knowledge, more security, etc.[42] This trend is closely aligned with transhumanist aspirations. For instance, in the *Letter from Utopia*, Bostrom discusses three fundamental transformations necessary to reach Utopia: a secure life, an expanded cognition, and an elevated well-being.[43] Furthermore, in *A Letter to Mother Nature: Amendments to the Human Constitution*, Max More emphasises the pursuit of "more", according to chosen values, as technology allows. Presented below are some extracts from More's work: "We will no longer be slaves to our genes. We will take charge over our genetic programming and achieve mastery over our biological, and neurological processes. We will fix all individual and species defects left over from evolution by natural selection."[44] More continues: "Not content with that, we will seek complete choice of our bodily form and function, refining and augmenting our physical and intellectual abilities beyond those of any human in history." Overall, the goal is not to achieve perfection: "rather

---

[41] Cf. NUFFIELD COUNCIL ON BIOETHICS, *Genome editing and human reproduction: Social and ethical issues*, Nuffield Council on Bioethics, London 2018, in https://www.nuffieldbioethics.org/publications/genome-editing-and-human-reproduction [20-4-2018], 23.

[42] P. D. HOPKINS, «A Salvation Paradox for Transhumanism: Saving You versus Saving You», in *Religion and Transhumanism: The Unknown Future of Human Enhancement*, ABC-CLIO, Santa Barbara 2014, 74–75.

[43] Cf. N. BOSTROM, «Letter from Utopia», *Studies in Ethics, Law, and Technology* 2/1 (2008), 1–7, in https://www.nickbostrom.com/utopia.html [12-7-2021], 3–5.

[44] «A Letter to Mother Nature: Amendments to the Human Constitution», 2009, in http://strategicphilosophy.blogspot.com/2009/05/its-about-ten-years-since-i-wrote.html [5-6-2020].

than seeking a state of final perfection, we will continue to pursue new forms of excellence according to our own values, and as technology allows."[45]

There is no doubt that "humans want to be happy, satisfied, safe and secure. Toward those ends, humans pursue love, food, shelter, community, power, status, relationships, wealth, health, respect, and beauty."[46] However, many people have all this yet still feel unfulfilled and find themselves desiring for more. Therefore, we should at least question the priority given to autonomy to fulfil desires over other values and principles.

HGE can certainly prevent and cure some diseases, yet human desire will steer HGE in a direction far beyond that. It is foreseeable that in the future HGE will, like other technologies, be at the service of the personal desires of individuals. Furthermore, the idea of HGE will likely be sold prior to the specific technology itself, pushing people to dream of what the product can offer. "The discoveries of genetics will not be imposed on us. Rather, they will be sold to us by the market as something we cannot live without."[47]

One might have thought that CRISPR, at least in the beginning, would be only used for curing rare genetic diseases. However, CRISPR is already being conceived for treating or preventing diseases on a wide scale. For example, following the success of CRISPR in lowering cholesterol levels in primates,[48] in July 2022 Verve Therapeutics launched a trial for a CRISPR-based therapy that alters the genetic code to permanently lower cholesterol levels. A woman in New Zealand became the first individual to receive this type of gene editing, permanently lowering her cholesterol levels.[49] This woman in particular had heart disease, along with an inherited risk for high cholesterol. However, researchers behind this clinical trial believe that this treatment could essentially help anyone interested in lowering cholesterol levels.[50] If the trials are successful, there is a potential that this specific use of CRISPR will soon become something offered to the general public as a one-shot vaccine. Lowering cholesterol levels in individuals will lead to preventing cardiovascular diseases, ameliorating health and well-being, and increasing life

[45] *Ibid.*

[46] P. D. Hopkins, «A Salvation Paradox for Transhumanism: Saving You versus Saving You», in *Religion and Transhumanism: The Unknown Future of Human Enhancement*, ABC-CLIO, Santa Barbara 2014, 74–75.

[47] Mark Frankel in F. Baylis, *Altered Inheritance: CRISPR and the Ethics of Human Genome Editing*, Harvard University Press, Cambridge 2019, 73.

[48] Cf. K. Musunuru et al., «In vivo CRISPR base editing of PCSK9 durably lowers cholesterol in primates», *Nature* 593/7859 (2021), 429–434, in https://www.nature.com/articles/s41586-021-03534-y [14-3-2023].

[49] Cf. Verve Therapeutics, «Verve Therapeutics Doses First Human with an Investigational In Vivo Base Editing Medicine, VERVE-101, as a Potential Treatment for Heterozygous Familial Hypercholesterolemia», 2022, *Verve Therapeutics*, in https://ir.vervetx.com/news-releases/news-release-details/verve-therapeutics-doses-first-human-investigational-vivo-base/ [3-2-2023]».

[50] Cf. J. Hamzelouarchive, «CRISPR for high cholesterol: 10 Breakthrough Technologies 2023», 2023, *MIT Technology Review*, in https://www.technologyreview.com/2023/01/09/1064857/crispr-high-cholesterol-10-breakthrough-technologies-2023/ [3-2-2023].

expectancy. "We know the basics of healthy living by now. A balanced diet, regular exercise, and stress reduction can help us avoid heart disease—the world's biggest killer. But what if you could take a vaccine, too? And not a typical vaccine—one shot that would alter your DNA to provide lifelong protection?"[51]

There is a potential for many good applications of HGE, and such applications are certainly desirable. However, technology at the service of desires becomes problematic in a framework where autonomy, understood as the right to accomplish personal aspirations, is the highest value. This concept will be further addressed in the coming chapter titled "Refining the Predominant Values for the Governance of Human Genome Editing".

To conclude this chapter, the commonalities between H+ and WHO texts regarding the impact of HGE technology create a task for Bioethics. Bioethics must expand its range of consideration to address the possibilities, desires, and questions related to:

- The increasing potential of HGE and how it might affect the human condition.
- The increasing availability of technology and the DIY mentality.
- Technology at the service of desires.

[51] J. HAMZELOUARCHIVE, «Next up for CRISPR: Gene editing for the masses?», 2023, *MIT Technology Review*, in https://www.technologyreview.com/2023/01/19/1067074/next-for-crispr/ [3-2-2023].

The Limitations of "Therapy (Allowed) Versus Enhancement (Forbidden)" 5

Abstract

A common theme that resulted from the Reflexive Thematic Analysis (RTA) method applied to selected WHO and Humanity + texts is *human genome editing for health, well-being, and enhancement.* Bioethics is tasked with constructing a better framework for the ethical analysis of interventions. This chapter highlights the limitations of the "therapy (allowed) versus enhancement (forbidden)" paradigm and moves towards defining a more coherent bioethical framework. Firstly, it considers the problem caused by equivocal concepts such as health, the goals of medicine, enhancement, and normalcy. Subsequently, this chapter analyses some aspects that weaken "therapy versus enhancement", such as the goal of disease prevention (claimed by both therapy and enhancement), the improvement in quality of life for the elderly (above the average for this age), and some commonly used enhancements. Lastly, this chapter acknowledges the weakest point of "therapy versus enhancement" as a criterion for morality. Ethical reasoning based exclusively on intentions lacks consistency and may cause significant harm in some cases while missing morally good or neutral improvements in other cases.

Keywords

Health • Therapy • Enhancement • Ethics • Medicine • Bioethics

L. Santos, *Governance of Human Gene Editing and Transhumanism*,
Integrated Science 28, https://doi.org/10.1007/978-3-031-70863-3_5

5.1 Health, Well-Being, Enhancement

A common theme that resulted from the RTA method applied to selected WHO and Humanity+ texts is *human genome editing for health, well-being, and enhancement.*[1] Throughout the selected WHO texts and transhumanist texts, the concepts of health and well-being were often presented as connected and regarded as priorities. As for enhancement, this was only considered a priority in H+ texts, and not in WHO texts. Since these three uses of HGE do not share the same level of importance within H+ and WHO texts, in the RTA, this theme was divided into priorities (health and well-being), and a possibility on the horizon (enhancement). Health, well-being, and enhancement are the three intended uses of HGE.

It goes beyond the scope of this work to present a literature review of the arguments regarding betterment. Therefore, this work limits itself to citing a summary that provides the arguments against and for enhancement. The reasons against it usually range from "playing God, bad character, contrary to human dignity/human rights, inauthenticity, harm to humanity, common heritage, human nature, social control, and inequality/justice."[2] The arguments for enhancement usually range from improvements "as not-so novel concern, inevitability, autonomy, benefit to others, obligation to better our children, and concerns that humans are unfit for the future."[3]

The reality is that enhancement has already arrived. There now exist various effective physical, cognitive, mood, cosmetic, and sexual enhancers. Enhancement "has moved from the realm of science fiction to that of practical ethics."[4] As a result, there is a need for bioethical consideration of enhancement. There is discussion concerning the possibility, as well as the utility, of differentiating between therapy and enhancement, especially if doing so will have a moral implication (such as allowing therapy and prohibiting enhancement). While discussions regarding definitions and moral classifications continue, innovations become available, and more are foreseen to arise. Amidst the uncertainty relating to the ethics of enhancement, the support for personal choice and self-determination is gaining popularity.

Bioethics is working with an unclear and changing framework regarding which interventions in the genome are to be ethically allowed. The limits of the "therapy versus enhancement" paradigm need to be recognised, and a better ethical framework needs to emerge. To consider the implications the weak paradigm has for

[1] The full Reflexive Thematic Analysis (RTA) can be consulted in the published doctoral dissertation excerpt L. SANTOS, *Common Themes Between Humanity+ Original Documents on Transhumanism and WHO Documents on the Global Governance of Human Genome Editing: Reflexive Thematic Analysis and Implications for Bioethics*, Ateneo Pontificio Regina Apostolorum, Rome 2023.

[2] G. EDWARDS, «Enhancement», in H. TEN HAVE (ed.), *Encyclopedia of Global Bioethics*, Springer International Publishing, Cham 2016, 1121–1130, 1123.

[3] *Ibid.*

[4] J. SAVULESCU - N. BOSTROM (edd.), *Human Enhancement*, OUP Oxford, Oxford 2009, 18.

Bioethics, it first should be noted that health and enhancement are two concepts not agreed upon.

5.2 The Equivocal Definitions Weakening the Framework

5.2.1 Health and the Goal(s) of Medicine

Health, the original goal of Medicine, is an extremely broad term. The definition of health, as stated in WHO Constitution (1946), has left the door open to the uncertainty that exists today between health and enhancement. "Health is a state of complete physical, mental and social well-being and not merely the absence of disease or infirmity. (…) The enjoyment of the highest attainable standard of health is one of the fundamental rights of every human being (…)."[5] If the concept of health is the complete state of well-being, could enhancement be considered as part of this desired state? Initially, the goal of medicine was defined as treating and preventing diseases, or in other words, maintaining and restoring health (the correct functioning of the organism). However, when the concept of health is broad as found in the WHO's definition, the goal of medicine becomes difficult to define.

Some people may wish to return to a more traditional and restricted meaning of health, in order to avoid such dilemmas and the related risks. However, this situation should be addressed by not only interpreting a concept but also interpreting recent experiences. Indeed, contemporary medicine embraces many actions that were not goals for doctors in the past. Today's standard of medicine includes many practices that do not aim to cure diseases or injuries. For example, preventive medicine, palliative care, obstetrics, sports medicine, plastic surgery, contraceptive devices, fertility treatments, cosmetic dental procedures, and many more.[6] There seems to be no sort of agreement regarding the true goal of medicine. Sandberg portrays this idea by pointing out that "We are rapidly approaching a time where there is not just curative, palliative and preventative medicine, but also augmentative medicine."[7]

If it was expected that the goals of medicine could indicate what are the ethical actions, today that is a problem. Many affirm that these goals are not static; they have changed over time, and have often come into conflict with one another.[8] "In

[5] WHO, «Constitution of the World Health Organization», 1946, in https://www.who.int/about/governance/constitution [29-11-2021], 1.

[6] Cf. N. Bostrom - R. Roache, «Ethical Issues in Human Enhancement», in J. Ryberg - T. Petersen - C. Wolf (edd.), *New Waves in Applied Ethics*, New Waves in Philosophy, Palgrave Macmillan UK, Hampshire and New York 2007, 120–152, in https://www.palgrave.com/gp/book/9780230537835 [12-7-2021], 120.

[7] A. Sandberg, *Morphological Freedom – Why We not just Want it, but Need it,* in The Transhumanist Reader, John Wiley & Sons, Ltd, Chichester 2013, in https://doi.org/10.1002/9781118555927.ch5 [14-4-2020], 56–64

[8] Cf. K. Fleischhauer - G. Hermeren, *Goals of Medicine in the Course of History and Today: A Study in the History and Philosophy of Medicine*, Stockholm 2006.

general terms, the goals of medicine include to combat disease, to alleviate suffering, to restore, maintain and improve health and quality of life. This is not one goal, but rather a cluster of non-equivalent goals."[9] This view provokes questions such as: Is enhancement the new and more adequate goal of medicine?[10] Or, could both therapy and enhancement be goals of medicine?

5.2.2 Enhancement

The word enhancement (and its synonyms: augmentation, improvement, or betterment) faces problems regarding definitions as well. For instance, "enhancement" has been implied to discuss a variety of issues, such as the use of hypoxic air machines for athletes, cosmetic surgery, off-label usage of medications, Ritalin usage to improve cognitive focus, oxytocin to improve social coherence, deep brain stimulation for tremor reduction and mood stimulation, memory erasures as a manner of coping with trauma, an exoskeleton for military and amputees, meditation, coffee, vaccinations, regenerative medicine, etc.[11]

The term enhancement is often used in opposition to interventions aimed at health, even when some improvements actually contribute to health. The concept of enhancement is also used to refer to any trait that is improved to increase human performance. In other occasions, enhancement is not so much connected to the objective improvement of any specific function, but to the acquisition of a desirable trait, considered as an improvement by the individual. Furthermore, enhancement and its synonyms can sometimes refer to the addition of a new trait (*new* can either mean beyond the individual's intrinsic capacity or beyond the species' capacity).

The term enhancement is often closely related to Transhumanism, and transhumanists are the main advocates of enhancement.[12] Nick Bostrom has defined enhancement as "An intervention that *improves the functioning* of some subsystem of an organism *beyond its reference state*; or that creates *an entirely new functioning* or subsystem that the organism previously lacked."[13] Beyond the transhumanist circle, the European Parliament's Science and Technology Options Assessment (STOA), states that enhancement definitions range from enhancement

[9] G. HERMERÉN, «The principle of proportionality revisited: interpretations and applications», *Medicine, Health Care and Philosophy* 15/4 (2012), 373–382, 374.

[10] Cf. P. BENANTI, «Il potenziamento cognitivo. Considerazione antropologiche ed etiche in prospettiva cristiana.», in G. QUARANTA (ed.), *Il doping della mente. Le sfide del potenziamento cognitivo farmacologico*, EMP, Padova 2014, 63–96, 64.

[11] Cf. G. EDWARDS, «Enhancement», in H. TEN HAVE (ed.), *Encyclopedia of Global Bioethics*, Springer International Publishing, Cham 2016, 1122–1123.

[12] Cf. HUMANITY+, «The Transhumanist Declaration», 2009, Humanity+, in https://www.humanityplus.org/the-transhumanist-declaration [11-6-2022].

[13] N. BOSTROM, «Dignity and Enhancement», in *Human Dignity and Bioethics: Essays Commissioned by the President's Council on Bioethics*, President's Council on Bioethics, Washington, D.C. 2008, 179. The italic is mine.

defined as *opposed to treatment*, enhancement as a *type of treatment* and *promotor of well-being*, enhancement as a relative concept *determined by the social dynamics*, and enhancement as a perception concept *determined by those affected* by it.[14] STOA's authors have declared that they do not rely on the widespread conceptual distinction between therapy and enhancement. Instead, they have adopted a notion of enhancement that includes non-therapeutic measures, as well as some therapeutic measures.[15] For "politically pragmatic reasons", STOA has defined enhancement as any "modification aimed at improving individual human performance and brought about by science-based or technology-based interventions in the human body."[16] However, STOA affirms that "embracing and tolerating the ambiguity of the concept "enhancement" may be necessary to some degree."[17]

Do the definitions matter? Ethical reflection will likely vary depending on the definition of health, therapy, prevention, and enhancement. "Inquiries into the morality of enhancement vary according to the defended definition, and such variety has arguably plagued the debate with diverse definitions (…)"[18] Bostrom and Roache argue that "defining the therapy-enhancement distinction is a problem only for those who maintain that this distinction has practical or normative significance."[19] Transhumanists are seemingly "unaffected by the problems associated with maintaining that there are important differences between enhancement and therapy."[20] However, if enhancement implies going beyond (individually or as a species) to improving it, what basis is this considered upon? What is the reference point from which to go beyond? What is the normal state to be improved upon?

5.2.3 Normalcy

How is normalcy defined? Is it equivalent to the average state? Or is it whatever state that nature has prescribed, or whatever luck has granted?"[21] Even if it has

[14] Cf. EUROPEAN PARLIAMENT'S SCIENCE AND TECHNOLOGY OPTIONS ASSESSMENT, «Human Enhancement. Panel for the Future of Science and Technology», 2009, in https://www.europarl.europa.eu/stoa/en/document/IPOL-JOIN_ET(2009)417483 [17–6-2021], 6.
[15] Cf. *Ibid.*
[16] *Ibid.*
[17] G. EDWARDS, «Enhancement», in H. TEN HAVE (ed.), *Encyclopedia of Global Bioethics*, Springer International Publishing, Cham 2016, 1121–1122.
[18] Cf. EUROPEAN PARLIAMENT'S SCIENCE AND TECHNOLOGY OPTIONS ASSESSMENT, «Human Enhancement. Panel for the Future of Science and Technology», 2009, in https://www.europarl.europa.eu/stoa/en/document/IPOL-JOIN_ET(2009)417483 [17-6-2021], 6.
[19] N. BOSTROM - R. ROACHE, «Ethical Issues in Human Enhancement», in J. RYBERG - T. PETERSEN - C. WOLF (edd.), New Waves in Applied Ethics, New Waves in Philosophy, Palgrave Macmillan UK, Hampshire and New York 2007, 122–123.
[20] *Ibid.*, 122.
[21] NATIONAL ACADEMIES OF SCIENCES, ENGINEERING, AND MEDICINE, «Human Genome Editing: Science, Ethics, and Governance», *The National Academies Press*, Washington, D.C. 14 February 2017, in https://doi.org/10.17226/24623 [5-7-2021], 138.

been agreed upon that the goal of medicine is to cure and prevent diseases, what constitutes a disease? In order to define disease, what is a normal state of health? According to NASEM, "Everyone would agree that the manifestation of Tay-Sachs disease is not normal and constitutes a disease, but opinions differ as to whether genetically caused deafness, dwarfism or autism should be considered a disease."[22] Perhaps deafness, dwarfism or autism are "variants that enrich human diversity"?[23]

Why is that so difficult to identify what constitutes normalcy nowadays? From a sociological point of view, Evans points out four problems concerned with the idea of normalcy in health. According to Evans, the problems are as follows:

- *Normalcy is based on the normative power of what currently is.*
  Notions such as nature, or God's design, are more likely to be found among the theologians who are not so present anymore in the public debate.
- *An objective definition of normalcy is in tension with individualism, where people have the right to define their own experiences.*
  In practice, the idea of "normalcy" assumes that one must be within a species-typical range to flourish. But there are some people with certain traits, such as deafness, who do not want to be changed. They do not believe they have a disease and consider themselves to be flourishing quite well. Therefore, a given objective measure does not necessarily match one's subjective interpretation of the measure.
- *Normalcy has become extremely susceptible to continuity vagueness.*
  Normalcy functions better alongside the concept of categorical disease—a disease that you either have or do not have and which manifests itself similarly across all those who have it, e.g. Sickle Cell Anemia. However, many genetic traits exist along a continuum. For instance, if one has a genetic propensity for heart disease, is a 50% increased prediction of heart disease normal? Is a 51% increase also normal? What about a 3% increase? What about the genetically influenced diseases that onset later in life, such as Alzheimer's at the age of 90, or 80…?
- *The very first HGE somatic therapy clinical trials in the U.S. hopelessly muddled the distinction between normality, disease, and enhancement in the debate.*
  The first human gene therapy trial, approved by the federal government in 1989, was not done because of an inherited disease. Instead, it constituted the genetic modification of normal immune cells in the body, to make them fight cancer more effectively.[24]

It is acknowledged the difficulty in determining what can be considered as normal in society today. What if the idea of normalcy migrates from the average to a self-defined state? For example, it is not average to be deaf, as most humans can hear.

---

[22] *Ibid.*, 148.

[23] *Ibid.*

[24] Cf. J. H. EVANS, *The Human Gene Editing Debate*, Oxford University Press, New York 2020, 62–64.

Yet, some deaf people do not consider themselves as having a disease that should be treated. Deafness can also be "associated with membership in a community of persons sharing this characteristic."[25] Is deafness a disease, or a less common but normal variant in the normal human?

What about "dis-enhancement", such as removing or worsening biological functions? To give a radical example, a pair of deaf parents undergoing in-vitro fertilisation may wish for their children to be deaf as well. After preimplantation diagnosis, they may decide to transfer only the respective embryos with the deafness tendency. With gene editing, it would be technically feasible to alter a gene so that deafness occurs.[26] This type of situation leads to "discussions about where the line is drawn between normal and pathological, as well as questions about who gets to draw these lines, what authority they have to draw them, which social dimensions are included or excluded, and what provision is made to contest the decisions."[27]

5.3 Some Confusing Scenarios

5.3.1 Prevention: Claimed by Both Sides

In addition to the problems associated with definitions, it is necessary to acknowledge the situation of prevention. Prevention is largely considered a grey area. It is viewed as a moral good and claimed by both sides (therapy and enhancement). For example, when a healthy person receives a vaccine, the shot improves the immunological system to prevent disease. Some may claim prevention relates to the therapeutic goal and so do vaccines. Others may claim that preventions such as vaccines are health-related but constitute an enhancement of the healthy and normal individual.

Another example is an action to lower cholesterol and prevent heart-related diseases. Today this is regularly done with medicines. In the future, doctors may have other methods to achieve prevention of heart-related diseases. In July 2022, Verve Therapeutics launched a trial for a CRISPR-based therapy that alters genetic code to permanently lower cholesterol levels. A New Zealand woman became the

---

[25] National Academies of Sciences, Engineering, and Medicine, «Human Genome Editing: Science, Ethics, and Governance», *The National Academies Press*, Washington, D.C. 14 February 2017, in https://doi.org/10.17226/24623 [5-7-2021], 148.

[26] Cf. European Group on Ethics in Science and New Technologies, *Opinion on the Ethics of Genome Editing*, Publications Office of the European Union, Brussels 2021, in http://op.europa.eu/en/web/eu-law-and-publications/publication-detail/-/publication/6d9879f7-8c55-11eb-b85c-01aa75ed71a1 [16–11-2021], 29–30.

[27] National Academies of Sciences, Engineering, and Medicine, «Human Genome Editing: Science, Ethics, and Governance», *The National Academies Press*, Washington, D.C. 14 February 2017, in https://doi.org/10.17226/24623 [5-7-2021], 148.

first to receive this gene editing.[28] In a scenario not so far in the future, we may see this type of gene editing offered to the general public as a one-shot vaccination. Lowering cholesterol levels prevents cardiovascular diseases, elevates health and well-being and increases the life expectancy of an individual. What if in addition to following a heart-healthy diet and increasing exercise, people could take a vaccine to prevent heart diseases? If a person with normal or average cholesterol for their age receives a treatment to lower their bad cholesterol levels and prevent heart disease, is this therapy or enhancement? Again, prevention is connected to the therapeutic goal, but in this scenario, the enhancement of a natural and normal trait occurs.

The distinction between therapy and enhancement has long been criticised by transhumanists. For instance, Bostrom and Roche reported six reasons as to why the therapy versus enhancement distinction is problematic:

- *The difference between standard-contemporary-medicine and medicine as it could be practised in the future.* Therapy-enhancement dichotomy is one thing, its relevance for the future of medicine is another.
- *It is unclear how to classify interventions that reduce the probability of disease and death.* E.g., vaccination can be seen as an enhancement of the immune system, or as a preventative therapeutic intervention that reduces the risk of contracting diseases. E.g., an intervention to slow the ageing process could be seen as an enhancement of healthspan, or as a preventative therapeutic intervention that reduces the risk of illness and disability.
- *It is difficult to define a "normal healthy state".* Many human attributes are inconsistently distributed across individuals. To define abnormality as deviations below the average population is to introduce an arbitrary point that seems to lack any medical or normative significance.
- *Human capacity varies continuously, not only within a population but also within the lifespan of a single individual.* As we grow old, our physical and mental capacities decline. If an intervention enables an old person to possess the same physical stamina, visual acuity, and reaction time as he had in his youth, does that constitute therapy or enhancement?
- *It is difficult to define how "internal" an intervention must be to count it as an enhancement versus a therapy.* For example, laser surgery is a therapy for poor vision. But what about contact lenses and glasses? What about a computer software that presents texts in an enlarged font?

[28] Cf. VERVE THERAPEUTICS, «Verve Therapeutics Doses First Human with an Investigational In Vivo Base Editing Medicine», VERVE-101, 2022, *Verve Therapeutics*, in https://ir.vervetx.com/news-releases/news-release-details/verve-therapeutics-doses-first-human-investigational-vivo-base/ [3-2-2023].

- *Is the moral status of enhancements different from that of other kinds of interventions that modify or increase human capacities to the same effect?* There seems to be a problem in justifying this claim.[29]

Outside transhumanist circles, Fr. Nicanor Austriaco has introduced the notion of "therapeutic enhancements",[30] regarding situations where individuals can be treated by enhancing something within them. This renders the moral contraposition between therapy and enhancement meaningless. For instance, Austriaco claims that the therapy-enhancement distinction should be replaced by a "therapy versus non-therapy distinction", which acknowledges that some therapies are enhancements.[31]

5.3.2 The Improvement of Health Conditions for the Elderly

Another grey area affecting the "therapy vs enhancement" framework is the improvement of health conditions for the elderly. These improvements help the individual whilst positioning him above average in health compared to their respective age group peers. According to WHO, "At the biological level, ageing results from the impact of the accumulation of a wide variety of molecular and cellular damage over time. This leads to a gradual decrease in physical and mental capacity, a growing risk of disease and ultimately death."[32] Common conditions experienced in this stage of life include hearing loss, cataracts and refractive errors, back and neck pain, osteoarthritis, chronic obstructive pulmonary disease, diabetes, depression and dementia. As we get older, the more we are likely to experience several of these conditions at the same time. Older age is also characterised by geriatric syndromes, including frailty, urinary incontinence, falls, delirium and pressure ulcers.[33]

Is the pain and suffering of the elderly population inevitable? If health is considered the average state of human capacities, what is the mitigation of the worsening of some capacities that naturally come with age (vision, hearing, joints, etc.)? It is common to gradually lose sight as one ages. Is it wrong to wear glasses? Is it wrong to get laser eye surgery? What if it were possible to prevent this loss of sight at the genetic level? Having good eyesight would be an above-average quality for elderly people, but does that make it wrong to intervene in improving

[29] Cf. N. Bostrom - R. Roache, «Ethical Issues in Human Enhancement», in J. Ryberg - T. Petersen - C. Wolf (edd.), New Waves in Applied Ethics, New Waves in Philosophy, Palgrave Macmillan UK, Hampshire and New York 2007, 120–122.

[30] N. P. G. Austriaco, «Healthier than Healthy: The Moral Case for Therapeutic Enhancement», *The National Catholic Bioethics Quarterly* 17/1 (2017), 43–49, in https://doi.org/10.5840/ncbq20171715 [5-7-2021].

[31] Cf. *Ibid.*, 49.

[32] WHO, «Ageing and health», 2022, *World Health Organization*, in https://www.who.int/news-room/fact-sheets/detail/ageing-and-health [27-2-2024].

[33] Cf. *Ibid.*

their quality of life? How far should we go to relieve their limitation, pain, and suffering? Are we doing enough to help them?

As Selvaraj points out, the rise of Transhumanism and the increasing number of secular individuals may lead to a significant change in people's perspectives on pain and suffering in the elderly. The advancement of technology, along with new ways of thinking and new expectations, has the potential to change the face of gerontology.[34]

The betterment of what is considered average for the elderly can be defined as therapy, as it is the restoration of a function that used to work well during younger stages of life. However, the very same action can be considered an enhancement of what is normal for an individual who falls within this age category. It is also worth noting that the improvement of a single trait may have positive results in other unrelated areas for the elderly. For example, improving sight may prevent falls and broken bones, and even increase life expectancy. Moreover, if treatments or enhancements for the elderly become widespread, they may end up modifying what is currently the average for this age group. In short, cases of improvement of quality of life for the elderly are another example in which therapy and enhancement are intertwined.

5.3.3 Some Enhancements in Use

To conclude this section concerning the specific scenarios that show "therapy allowed, enhancement forbidden" as a simplistic and weak paradigm follows a review of some enhancements already being used in society.

Examples of when the objective is creating an improvement for health purposes include the use of vaccinations, vitamins, supplements, and the lowering of normal cholesterol levels to prevent heart disease.

Another example is in the area related to the use of prosthetic devices. A prosthesis is defined as any given artefact used to restore bodily functions. Health and enhancement are closely connected regarding the way that prostheses are developing. With the aid of modern prostheses, many human biological functions can be improved and restored. Not only that. Some external and internal prosthetic artefacts have more features or perform better than the natural ones that they are replacing.[35]

In terms of improving the quality of life for the elderly population, some common options include using hearing devices, wearing eyeglasses and receiving laser surgery. As people age, it is natural for eyesight and hearing to decrease in quality, thus necessitating the use of these devices.

---

[34] Cf. F. J. SELVARAJ, «Transhumanism and improving quality of life in the elderly: How could it shape the future of pain and suffering in the elderly?», *Medicina y Ética* 35/1 (2024), 61–107, in https://doi.org/10.36105/mye.2024v35n1.02 [27-2-2024].

[35] Cf. P. BREY - S. NAGEL, «Bioengeenering», in H. TEN HAVE (ed.), *Encyclopedia of Global Bioethics*, Springer International Publishing, Cham 2016, 280–291, 284–285.

In addition to the previous examples, there are also enhancements which cater to personal preferences, such as hair dyes, teeth whiteners, etc.

In this sense, the WHO framework for the global governance of HGE demands "attention to the full range of temporary and permanent medical and biological enhancements already in use, ranging from nutritional supplements to strength-enhancing steroids to cosmetic surgeries."[36]

Should we reject therapeutic enhancements? Should we reject enhancements based on aesthetic or personal preference reasons? Should we oppose any sort of improvement of what is considered normal or average? Are all forms of improvement intrinsically bad? Altogether, a framework that considers all enhancements as morally wrong, solely because some specific enhancements may be harmful, is difficult to consistently uphold.

5.4 The Flaws of Ethical Reasoning Centred on the Intention

There is a final and crucial element regarding the limitation of the framework "therapy-right, enhancement-wrong". Even in cases where the intention is clear as being either therapy or enhancement, this does not imply that the given case is morally acceptable. The intention behind the action is not the only factor in determining its ethical standing. Indeed, not everything proposed for therapy purposes is morally good, simply because it is based on the virtuous intention to restore health (e.g., consider situations involving organ trafficking, forced therapies, or futile treatments). Conversely, not everything proposed for improvement purposes is immoral (e.g., enhancing normal cells to prevent or fight cancer, playing classical music to an unborn baby to improve chances of musical talent, using glasses when it is natural to lose sight at an elderly age).

WHO recognises that gene editing may one day be used to improve typical and perfectly adequate traits, or add new traits, or increase the so-called desirable traits.[37] This possibility raises many concerns, yet the same WHO urges us to pay "attention to the full range of temporary and permanent medical and biological enhancements already in use".[38] It does not seem fair to completely oppose all enhancements due to the concerns associated with some of them. Clearly, there are serious concerns surrounding certain types, intentions, and circumstances of enhancement that need to be addressed. In Chap. 6, this book will provide a deeper analysis of these primary concerns.

[36] WHO Expert Advisory Committee on Developing Global Standards for Governance and Oversight of Human Genome Editing, *Human genome editing: a framework for governance*, WHO, Geneva 2021, in https://www.who.int/publications/i/item/9789240030060 [13-7-2021], 26.

[37] Cf. *Ibid.*

[38] *Ibid.*

Understandably, some individuals might choose to defend the "therapy allowed, enhancement forbidden" paradigm as a means to avoid the dangers surrounding certain enhancements. This polarisation may be motivated by the fear of unlimited liberty, which can indeed cause great harm to the individual and to society. A position which defends unlimited freedom is essentially one that accepts harm. For example, such a position would accept the act of suicide as long it was voluntary.

In all honestly, drawing the moral line between therapy and enhancement seems to be an easier, yet weaker solution against unlimited liberty. Although it feels safer to defend only therapy-related interventions, for all the reasons explained earlier, this is not an intellectually honest option to me. There is not sufficient rationality in basing morality exclusively on intention. Furthermore, creating improvements is not always necessarily a bad goal. Additionally, there is a disadvantage in holding onto the safer paradigm: by opposing all enhancements to avoid the potentially dangerous improvements, we would also miss out on the good and neutral applications.

Furthermore, total opposition to enhancement contradicts actions that have aimed to create improvements throughout human history. Human history portrays different forms of good and neutral human interference to the natural state for various purposes such as survival, the restoration of health, prevention, improvements, and modifications according to preference.

To conclude this chapter, there is a great deal of discussion surrounding the possibility (and the utility) of differentiating between therapy and enhancement, especially if this division will have a moral connotation such as allowing therapy and prohibiting enhancement. Some equivocal concepts and some confusing scenarios evidence that "therapy allowed, enhancement forbidden" is a falling paradigm. Amidst this unclarity, personal choice and self-definition are advancing. While there continues to be plenty of debate regarding the definitions and moral classifications for therapy and enhancement, new interventions are becoming available, and more are foreseen in the near future.

From "Therapy Versus Enhancement" to Proportionality

6

Abstract

Recognising the limitations of a framework that allows therapy and prohibits enhancement based on intention, this chapter questions whether enhancement (understood as an intervention to improve a human trait above the average level in a population) is intrinsically bad. An analysis of the object, intentions, and circumstances of enhancement was done to verify that. The intentions can be divided into a *proximate* end (improvement) and some *remote* deeper intentions (e.g., to cure or prevent diseases, to be superior to others, to be fit for a specific activity, etc.). The circumstances surrounding enhancements are summarised as the risk of discrimination, social rejection of human vulnerability, inequality of access, loss of diversity, and the lack of liberty. However, if all the above were solved, at least to a satisfactory level, the question of whether or not enhancement is inherently wrong remains. It is necessary to analyse the object of enhancement, which may be neutral, good, or bad for the person and the species. This depends on the specific intervention and the means involved. The means must be always legitimate (good or neutral) and adequate for the specific case. Finally, this chapter initiates a reflection on a new framework that is comprised of the ethical assessment of the object, intentions, and circumstances of enhancements and calls for the principle of proportionality to guide the selection among legitimate means.

Keywords

Health • Therapy • Enhancement • Ethics • Proportionality • Transhumanism

L. Santos, *Governance of Human Gene Editing and Transhumanism*,
Integrated Science 28, https://doi.org/10.1007/978-3-031-70863-3_6

6.1 Assessment of the Object, Intentions, and Circumstances of Enhancement

In order to offer a more elaborate framework than "therapy allowed, enhancement forbidden", it should first be determined whether or not enhancement is intrinsically bad. Some individuals, such as bioethicist John Harris, argue that enhancement is intrinsically good. "If it wasn't good for you it wouldn't be enhancement."[1] With these words, Harris claims that enhancement is good by definition. But I agree with M. Hall who says that "one cannot assume that enhancement is morally acceptable without, in effect, refusing to enter the debate before it begins."[2] The moral question regarding enhancement is pertinent and must be undertaken, even in John Harris' view. For instance, if moral goodness was inserted into the definition of enhancement, the matter would simply switch to "if enhancements are morally good, *which* interventions can be considered enhancement?" Therefore, one way or the other, interventions for the purpose of improvement require bioethical reflection.

Many alterations to what is naturally present are carried out for distinct purposes. Examples include human interventions for the preservation of life and restoration of health (e.g. xenotransplants, pacemakers, gene therapies, implants), for the improvement of health (e.g. vitamins, supplements), for the prevention of diseases (e.g. vaccines), for psychological well-being (e.g. certain plastic surgeries), and even for beauty purposes (e.g. cosmetics). As can be observed, these interventions vary by their intentions and the means involved.

Since the differentiation between enhancement versus therapeutic intention is insufficient to determine morality, an analysis of the object, intentions, and circumstances of the intervention should take place. Using this framework as a starting point means that individuals who follow other ethical frameworks, such as Utilitarianism, Relativism, or consensus-based frameworks, may find it difficult to agree with various aspects of my analysis. These differences could be summarised by the evil that these frameworks are willing to accept in order to achieve different goals. For instance, Utilitarianism accepts evil to obtain pleasure for a higher number of people. On the other hand, Relativism is not concerned with objective moral evil, since good versus evil is defined by each person, and individuals should pursue their self-defined good even if this involves using evil means. Lastly, contractualist positions defend the notion that the important matter is that parties have agreed on an issue. However, those who are willing to avoid evil in the object, intentions, and circumstances of their actions, will find the upcoming ethical analysis to be a useful tool.

According to the classical ethics deliberation framework, human actions are morally good when the object, ends (intentions), and circumstances are good. The

[1] J. Harris, «Enhancements are a Moral Obligation», in J. Savulescu - N. Bostrom (edd.), *Human Enhancement*, OUP Oxford, Oxford 2011, 131–154, 131.

[2] M. Hall, *The Bioethics of Enhancement: Transhumanism, Disability, and Biopolitics*, Lexington Books, Lanham and London 2016, 3.

object, the ends and the circumstances of an action are called the "moral sources". Each of these three sources is chosen by the free will of the moral agent, so all must be good for the action to be good.[3] The "do good, avoid evil" notion is the most generalised compass that can be followed when aiming to engage in moral actions. If the aim is to do good and avoid evil, then moral evil should be rejected within the three moral sources. It implies that when we accept evil, either in the object, intentions or circumstances of the action, we are already engaging in evil, which is what we had been aiming to avoid.

A different case is the principle of double effect[4] in which there are bad, potentially foreseeable, and undesired consequences associated with a given action. However, it should be noted that a bad consequence is different from an evil object or an evil means. Particularly when it relates to bioethical issues, it is important to remember that when evil is accepted within either the object, intentions, or circumstances of the action, individuals are treated unjustly and their lives, integrity, and health may be harmed.

Firstly, I will explore whether enhancement, defined as a human intervention to alter an average trait to improve its performance, always constitutes a morally wrong act. If so, there is no sense in analysing the specific enhancement proposals. However, if enhancement is not intrinsically bad, then some enhancements may be morally good, while others may be morally wrong, depending on other factors. In this case, the challenge would be to determine the conditions for the morality good enhancements. Then, with the help of these proposed criteria, ethical analysis would need to be done on a case-by-case basis.

---

[3] Austriaco explains the morality of acts by citing Thomas Aquinas' *Summa Theologiae* I-II, q. 18, a. 4, ad 3; I-II, q. 19, a. 6. For an act to be good, every moral source—the object, the intention, and the circumstances—has to be good. Each moral source is chosen by the will, so each must be good for the will to be good. The scholastic axiom—*malum ex quocumque defectu*, or "evil comes from a single defect"—encapsulates this moral truth that the whole act is evil if even one of the moral sources of an act is not in accord with right reason, that is oriented to the authentic good of the person and community. This is because any evil in the act would disorder the will, making the will evil. Cf. N. P. G. AUSTRIACO, *Biomedicine and Beatitude*: *An Introduction to Catholic Bioethics*, The Catholic University of America Press, Washington, D.C. 2021(2), 29.

[4] For further reading on the principle of double effect, Cf. J. MANGAN, «An Historical Analysis of the Principle of Double Effect», *Theological Studies* 10 (1949), 41–61, in https://doi.org/10.1177/004056394901000102 [16-3-2023]; Cf. BOYLE JOSEPH M., «Toward Understanding the Principle of Double Effect», *Ethics* 90/4 (1980), 527–538, in https://www.journals.uchicago.edu/doi/abs/10.1086/292183 [16-3-2023].

It is important to note that some conditions must be met if the agent aims a good action with double effect: the object must be morally good or neutral, the intention should be directed towards the beneficial effects and not to the foreseen bad effects, the beneficial effect must not be the result of the evil effect, and the beneficial effect must be equal or greater than the foreseen harmful effect. In summary, in the moral order, good must be proportionate to the bad effect. Cf. N. P. G. AUSTRIACO, *Biomedicine and Beatitude: An Introduction to Catholic Bioethics*, The Catholic University of America Press, Washington, D.C. 2021(2), 33–35.

Therefore, what follows is a general screening which reviews whether an enhancement is intrinsically good or bad and merits consideration in a more concise bioethical analysis. To clarify, enhancement in this analysis is understood as a human intervention that improves a human trait within the current average parameters of a population.

6.1.1 Intentions: Improvement and Possible Deeper Intentions

The intention or end is the motivation of a human act, the purpose that prompts someone to engage in a certain action. Every human action is performed on the basis of intention, and often these intentions have varying levels. For example, regarding the case of enhancement, the *proximate end* is to improve a human trait. This is the closest and most superficial intention of this kind of intervention. *Remote ends* are deeper and can vary widely. For example, an improvement to cure an illness, to prevent disease, to be superior than others, or to increase fitness to excel in a certain activity, a specific mission, work, sport, etc. Altogether, the intention behind a given action influences its morality. As a result, a negative motivation behind an enhancement results in the action being morally wrong.

It is worth noting that transhumanists emphasise the potential good intentions of enhancement: "(...) we should seek to develop and make available human enhancement options in the same way and for the same reasons that we try to develop and make available options for therapeutic medical treatments: in order to protect and expand life, health, cognition, emotional well-being, and other states or attributes that individuals may desire in order to improve their lives."[5] The betterment intention is a common argumentative strategy, used predominantly by the supporters of enhancement, to highlight the commonalities between the new controversial methods of enhancement and the older accepted ways of enhancing human capacities.[6]

For example, Bostrom and Savulescu ask: How is taking modafinil fundamentally different from imbibing a good cup of tea? How is it morally different from getting a full night's sleep? Are not shoes a kind of foot enhancement, clothes an enhancement of our skin? A notebook or a smartphone can be viewed as a memory enhancement. In short, all technology can be viewed as an enhancement of our native human capacities, enabling us to achieve certain effects that would otherwise require more effort or be altogether beyond our power.[7]

---

[5] N. Bostrom - R. Roache, «Ethical Issues in Human Enhancement», in J. Ryberg - T. Petersen - C. Wolf (edd.), New Waves in Applied Ethics, New Waves in Philosophy, Palgrave Macmillan UK, Hampshire and New York 2007, in https://www.palgrave.com/gp/book/9780230537835 [12-7-2021], 122–123.

[6] Cf. N. Bostrom - J. Savulescu, «Human Enhancement Ethics: The State of the Debate», 2009, in https://www.nickbostrom.com/ethics/human-enhancement-ethics.pdf, 2.

[7] *Ibid.*

Indeed, improvement as a proximate end is good. But the remote and deeper intentions should be considered as well. If the intention is considered good, the act has the potential to be morally right, but that is not guaranteed. Moral goodness also depends on the object and circumstances of the action. Saying that the intentions of enhancement are good is different to saying that all situations and methods intended for enhancement are good. The good intention behind enhancement is not enough to constitute a good action. The intention is not the only factor in condemning enhancement, and it should not be the only factor in accepting it.

As previously mentioned, moral legitimacy does not reside in the idea that therapy is ethical (intention of curing and preventing disease) and enhancement is unethical (intention of improving a normal human function). For instance, although therapeutic intention is generally regarded as ethical, this does not ensure that every therapeutic act is morally good. Organ transplantation to save a life is an intervention with laudable intention, however, organ trafficking can completely change the morality of this action. In the case of enhancement, improvement above the average level is not inherently bad. For example, athletes can enhance the functioning of their bodies above average through diet and training, and vaccines can enhance our immune system above average to prevent diseases. Some enhancements are desirable, good, and worthy to improve either health, the quality of life or individual preferences. The more is not always the better, but it does not mean the more is necessarily the worse.

Most people who oppose human enhancement do not base their opinion on the intention of enhancement. Improvement, the proximate end, is a good thing. Although certain remote intentions may be bad, there is no reason to say all remote intentions are always bad. It is more reasonable to believe that the concerns reside in the means used for these improvements and the circumstances of the act. Although the intention is not always bad, enhancement is frequently surrounded by various risks.

6.1.2 Circumstances: Coercion? Discrimination? Inequality? Uniformity?

The circumstances include the particulars of a specific action which may affect its morality, such as time, persons involved, occasion, etc. The circumstances can make a good action better and an evil action worse. All human actions occur in a particular context, and therefore the given circumstances must always be considered in evaluating the morality of an action.

Most concerns regarding enhancement are associated with the circumstances surrounding the intervention. The most typical include:

- Discrimination.
- The social rejection of human vulnerability and disability.
- Inequality of access to enhancement tools.
- Too much equality and uniformity resulting in the loss of diversity.

- Lack of liberty.
  - Coercion, either explicit or implicit.
  - External influences on desires and choices, the idea of perfection, etc.
  - Cases of decisions made by others (e.g. decisions made for future generations and for people who are unable to consent).

These concerns should not be ignored. Although they are not part of the object, nor the end of enhancement, there are significant risks related to these circumstances.

However, a valid question to ask at this point is: What would happen if the circumstantial concerns regarding enhancement were resolved or reached a satisfactory level? Circumstantial concerns are the ones that do not address the fact of enhancement in itself (which is an essentialist concern), but rather the issues surrounding enhancement, such as the examples mentioned above. Differentiating circumstantial from essentialist concerns does not mean that circumstantial concerns are not important. It rather shows that if circumstantial concerns could be prevented or resolved, the question about the ethical acceptability of enhancement interventions would still be there. The most important question that should be answered is: Is enhancement a problem in itself? Or does the problem lie only in circumstantial issues such as possible inequality of access, the tendency towards social discrimination, the factors limiting real freedom, and so on?

> Let us suppose that the questions of safety—mishaps, creation of monsters, and unpredictable effects on the human gene pool—eventually disappear as technology overcomes these obstacles. Let us also suppose that the question of justice—Nazi-style eugenics, social discrimination, and the unfair distribution of this technology—can be eliminated by affordability, availability, and governmental policy to ensure fairness. The question that still remains is whether genetic perfection is absolutely desirable or would impoverish us, paradoxically making us less human?[8]

Most of the objections to enhancement seem more for cautious reasons than a justification for a prohibition motivated by non-controversial reasons.[9] Although these concerns are reasons to be cautious, they do not constitute a final case against enhancement, at least not against all kinds of enhancements. What would happen if technical and social concerns were resolved or sufficiently mitigated? Having checked the possibility of good intention and of good circumstances for enhancement, one factor still has to be analysed: the object.

---

[8] J. THAM, «Resisting the Temptation of Perfection», *The National Catholic Bioethics Quarterly* 17/1 (2017), 51–62, 53–54.

[9] Cf. M. REICHLIN, «Etica delle neuroscienze: il caso del potenziamento cognitivo.», in G. QUARANTA (ed.), *Il doping della mente. Le sfide del potenziamento cognitivo farmacologico*, EMP, Padova 2014, 19–62, 58.

6.1.3 Object: The Specific Intervention and the Choice of Means

The object of human action is *what* an individual with his free will chooses to do. It is generally classified as morally good, bad, or indifferent. It is usually the primary factor for moral judgment. However, in the present analysis, I decided to address first the intention and circumstances and then the object. This was because the usual framework opposes therapy-intended interventions to enhancement-intended ones as a moral criterion. Therefore, the intention is given high importance in the discussion. Then, most of the cited concerns refer to circumstances surrounding enhancement interventions and for this reason, I have also analysed the circumstances before moving to the object.

Considering enhancement in general, the object is the intervention to improve some aspect of human performance above the average level present in an age group or a population at the current time. This object might be neutral, good, or bad for the person and the species. It will depend on the specific intervention that increases human performance. The distinction should be made on a case-by-case basis.

In the analysis of the object, due regard is to be given to the means employed in the act. First, the legitimacy of the means is to be assured: the means must be good or neutral. No good intention or good circumstances can make good an evil object or justify the use of evil means. Otherwise, there is no sense in aiming at moral good if it implies moral evil. For instance, research subjects must be protected and only be enrolled in the research with corresponding informed consent. People enrolled in clinical trials can contribute to science, even if the unknown and unwanted side effects result in bad consequences. However, these people should never be sacrificed nor exploited even for a good purpose like the progress of science. When this happened in history, even with the acceptance of the authorities, it constituted nevertheless a moral evil and had devastating consequences for the people involved. In short, first, the means should be *legitimate* (good or neutral), and second, the means should be *adequate* for the specific case. A means indicated for a case might not be adequate for another case.

In the case of interventions, an important question is: what about the means that carry health risks? Risky means are not moral evils, but they depend on other factors to be appropriate in a specific case, e.g. the level of necessity, and the risks versus benefits balance. Risky means should be proportionally justified by the need and the potential benefits, to respect the "do not harm" imperative.[10] What is the problem with disproportional means? It fails the risks-to-benefits ratio, causing

[10] Burden-benefit approach has a long history in medical intervention. The risk balance is necessary to respect the modern bioethical principle of non-maleficence. Cf. T. L. BEAUCHAMP - J. F. CHILDRESS, *Principles of Biomedical Ethics*, Oxford University Press, New York 2001[5]. The famous "first do not harm", *primum non nocere*, is belived to date from the seventeenth century. But it has its roots in the Hippocratic school. In Epidemics, Book I, is written: "Practice two things in your dealings with disease: either help or do not harm the patient." HIPPOCRATES, *Hippocratic*

harm that could have been and should have been avoided. Putting a person at unnecessary risk by using disproportionate means, will not make this means (e.g., genetic intervention) wrong on all occasions, but will constitute a moral evil in some cases. Why? Because the use of a means that puts a person at risk of harm without a proportional necessity is unjustified and irresponsible.

To illustrate the consequences of this, let us examine an example. Bostrom complains about the existence of "one standard for germ-line genetic interventions and another for improvements in maternal nutrition (an intervention presumably not seen as heralding a posthuman era)." He gives the example of an accepted enhancement: playing Mozart to an unborn child to increase the chances of the child developing musical talent. Bostrom compares it to genetic enhancements and complains that some people would accept the first and reject the second. His exact words were:

> Suppose it turned out that playing Mozart to pregnant mothers improved the child's subsequent musical talent. Nobody would argue for a ban on Mozart-in-the-womb on grounds that we cannot rule out that some psychological woe might befall the child once she discovers that her facility with the violin had been prenatally 'programmed' by her parents. Yet when, for example, it comes to genetic enhancements, eminent bioconservative writers often put forward arguments that are not so very different from this parody as weighty, if not conclusive, objections. To transhumanists, this looks like doublethink.[11]

Why does Bostrom not consider that playing music and genetic interventions are very different means for the same end? The controversies are not due to the improvement *intention*, but rather to the different *means* proposed. And there is not even a problem with the mean itself (gene editing), but the appropriateness of its use in this case. Although the intention of improving musical ability is the same, the means make the cases quite different. In a simple risk versus benefit assessment approach, the outcome is clear that playing music to the fetus in the womb has the potential for benefits while presenting a low or inexistent risk for the mother and baby. In contrast, for example, somatic (*in uterus*) genetic intervention would carry health risks for the child and the mother. Risks that are not worth taking to increase the chances of being a talented musician. A good intention does not justify the use of all possible means to achieve the desired goal. The means must be, first, legitimate, that is, good or neutral. Second, the means need to be adequate, that is, proportional to the need and considering the risk versus benefit balance.

In the case proposed by Bostrom, the method (gene editing) is not inherently bad. *In uterus* somatic gene editing is not an intrinsically bad method. It could be a type of more precise *in uterus* surgery that could save or cure the unborn child. The confirmation that it is a disproportional and therefore an inappropriate method

---

Writings, edd. J. CHADWICK - G. E. R. LLOYD - W. N. MANN, Penguin classics, Penguin Books, Harmondswerth 1983, 94.

[11] *Ibid.*, 212.

for one scenario does not make *in uterus* HGE an option to discard use in all other scenarios. We can theorise that *in uterus* somatic HGE might be an adequate means to treat a genetic disorder, and we can hope that it will become a safer and more available option in the future. However, in the case presented by Bostrom, either heritable or somatic (*in uterus*) HGE would put the mother and child at risk, which is not justified for the sake of increasing musical ability. It should be noted that there is nothing wrong with the intention either. Improvement of musical talent is not bad, but it should be pursued with less dangerous means.

The above example shows the following: "Comparisons with the development of persons sought through physical and spiritual means, education, mentoring, etc. are misleading."[12] The logic implied by transhumanists to compare enhancements comprehending done by low-risk means to enhancements by riskier means completely fails. That is because "regardless of the similarity of the intent, the actual nature of the action makes them significantly different."[13]

In other words, the element that seems to be lacking in many of the transhumanist proposals, in my point of view, is proportionality to select the right means among the legitimate ones. Not all kinds of means can be employed to pursue a good goal. When some considerable risk is involved, the choice of adequate means depends on many factors. For instance: What is the level of necessity for the use of risky means? What are the risks and potential benefits? Is this specific means recommendable for this person, considering, for example, his health conditions? Are there other means available?

Indeed, when talking about enhancements, comparing situations that have the same intention to justify the use of all sorts of means can lead to dangerous situations. However, Bostrom has a point criticising bioconservatives in the previously quoted text. There is a tendency to contrast transhumanist proposals either by emphasising circumstantial concerns or by fighting against the intention of improvement. However, it seems that the most appropriate reasoning against some specific enhancements has to do with the disproportionate use of means.

6.2 Ethical Conditions for Enhancement by Non-transhumanists

If enhancement does not constitute evil always and a priori, it should be evaluated on a case-by-case basis. Some authors who follow frameworks very distant from Transhumanism have also reached this conclusion. For instance, bioethicist Postigo says "That is my general assessment of human betterment. Not everything as a whole is immoral, not everything is good. Judgement must be made on an

[12] G. MENDZ - M. COOK, «Transhumanist Genetic Enhancement: Creation of a 'New Man' Through Technological Innovation», *The New Bioethics: a multidisciplinary journal of biotechnology and the body* 27/2 (2021), 105–126, 123.

[13] *Ibid.*, 116.

intervention-by-intervention basis."[14] If enhancement is not an intrinsic evil, what are the conditions for a morally acceptable enhancement? What are the criteria to evaluate each enhancement proposal? There still needs to be further development in Bioethics. The following are some attempts at this task.

Understanding enhancement in a more holistic way- "(…) not only the improvement of physical capacities, but a holistic improvement of the person, the attainment of personal fulfilment in a global sense, the accomplished life of the individual",[15] Postigo affirms that we should consider, one by one and in detail: what each intervention means, what they imply, the intentions and goals, and what means would be used.[16]According to Postigo, we should be open to possible interventions that:

- Respect the do not harm principle (*primum non nocere*).
- Serves human progress and the common good.
- Do not violate the fundamental rights or fundamental norms of ethics and human life in society.[17]

Another Spaniard scholar, Cortina, says scientific progress is legitimate in terms of alleviating human suffering (the eradication of diseases, the prolongation of well-being for all, etc.), as well as the promotion of certain human capabilities.[18] The conditions he proposes for morally acceptable enhancements are that the interventions:

- Do not compromise other goods and values
- Are not imposed coercively.
- Do not involve risks greater than the potential benefits.[19]

A third proposal comes from Fr. Nicanor Austriaco. Instead of therapy *versus* enhancement, he proposes that in some cases individuals are treated *by* enhancement. In his words: "I therefore propose that a robust argument can be made for the moral liceity of genetic interventions that treat by enhancing. They make patients

---

[14] E. POSTIGO, «Naturaleza Humana y Problemas Bioéticos del Transhumanismo y el Mejoramiento Humano», in A. CORTINA - M.-À. SERRA BELTRÁN (edd.), *Humanidad infinita: Desafíos éticos de las tecnologías emergentes*, EIUNSA, Madrid 2016, 231–256, 244. The translation is mine. Follows the original: 'Esa es mi apreciación general respecto al mejoramiento humano. Ni todo en su conjunto es inmoral, ni todo es bueno. Se ha de juzgar intervención por intervención.'.

[15] *Ibid.* The translation is mine. Follows the original: '(…) por mejoramiento humano entendemos no solo la mejora de sus capacidades físicas, sino un perfeccionamiento integral de la persona, llegar a la plenitud personal en sentido global, a la vida lograda.'

[16] Cf. E. POSTIGO, «Bioética y transhumanismo desde la perspectiva de la naturaleza humana», *Arbor* 195/792 (2019), a507, 9.

[17] Cf. *Ibid.*

[18] Cf. A. CORTINA, *Humanismo avanzado para una sociedad biotecnológica*, EIUNSA, Madrid 2017, 65.

[19] Cf. *Ibid.*

healthier than healthy."[20] Austriaco works with a framework where the goals of medicine are to restore, preserve and protect health. "I would argue that medicine does more than simply restore the patient to its normal state or counter damage caused by pathology or anomaly. Medicine seeks not only to restore, but to preserve and protect the health of individuals."[21] This framework justifies vaccines, which are applied to healthy individuals to improve the immune system to prevent diseases. Austriaco also says this framework morally permits interventions to decrease LDL or "bad" cholesterol levels lower than average in our species to prevent cardiovascular diseases.

> How else could we justify vaccination programs like the annual flu shot, which can make otherwise healthy individuals feel unwell in the short term in order to prevent serious diseases in the long term? For the same reason, I propose that it should be morally permissible to significantly decrease LDL to levels that lie outside what is normal or typical for our species in order to protect the health of those who may die from cardiovascular disease.[22]

Austriaco states: "To put it more generally, it should be morally justifiable to enhance our biological capacities for therapeutic reasons."[23] Specifically applying it to HGE, he says genetic interventions, including those that improve and strengthen the human genome pool, are morally permissible, "as long they seek to preserve and protect the health of those individuals whose genomes are edited."[24] His interpretation of the therapeutic goal concludes that HGE should also be used to protect and preserve human health. "These would justify genetic interventions that improve and strengthen the gene pool beyond species norms as long as they are undertaken for therapeutic reasons. Gene editing to make patients healthier than healthy should be allowed."[25]

Austriaco is not the only author who moved from the binary distinction between therapy and enhancements, with consequences to the applications of HGE. Some talk about genetic protection treatments: "(…) genetic treatment of sporadic Alzheimer's disease, breast and ovarian cancer predisposing BRCA1/2 mutations, and the introduction of human immunodeficiency virus resistance in humans should be considered within a new category of genetic protection treatments."[26]

---

[20] N. P. G. AUSTRIACO, «Healthier than Healthy: The Moral Case for Therapeutic Enhancement», *The National Catholic Bioethics Quarterly* 17/1 (2017), in https://doi.org/10.5840/ncbq20171715 [5-7-2021], 46.

[21] *Ibid.*, 47.

[22] *Ibid.*, 47–48.

[23] *Ibid.*

[24] *Ibid.*, 48.

[25] *Ibid.*, 49.

[26] R. B. MIKKELSEN ET AL., «Genetic Protection Modifications: Moving Beyond the Binary Distinction Between Therapy and Enhancement for Human Genome Editing», *The CRISPR Journal* 2/6 (2019), 362–369, in https://www.liebertpub.com/doi/abs/10.1089/crispr.2019.0024 [4-3-2021], 362.

In Austriaco's proposal, "the therapy-enhancement distinction should be replaced by a *therapy-nontherapy* distinction that acknowledges that *some therapies are enhancements*."[27] He also mentions that "the *benefits-burdens distinction* should be employed, as it commonly is with other technologies that affect the human person, alongside the therapy-nontherapy distinction to judge whether a particular genomic intervention should be permitted or not."[28] Specifically, HGE to reduce the risk of cardiovascular disease, immunise against HIV infection, or prevent the development of cancer, "could only be justified if the benefits of such an intervention outweigh any accompanying burdens. Persons can only be made healthier than healthy if they do not unreasonably put their health at risk."[29] In a few words, Austriaco's therapy-nontherapy paradigm recognises some enhancements for therapeutic reasons (those aimed at health restoration, preservation, and protection). And he mentions the importance of the benefits-burdens distinction, without developing the idea in cited article.

Agreeing with Austricio's framework, Chap. 7 of this book tries to develop the benefit-burdens distinction to choose the adequate means for interventions. Risk versus benefit assessment, measured on probabilities, is often used in medical decision cases.[30] However, the principle of proportionality offered in the next chapter is proposed for interventions beyond the medical field. Chapter 7 expands the reflection beyond health to other good or neutral purposes.

6.3 Towards Proportionality

Leaving behind *therapy vs enhancement* as the criterion for morality, ethical reasoning should move towards proportionality to choose among legitimate means for interventions. Disproportional use of risky means can make the action dangerous, not recommended, and unethical. Modifications and interventions in the body are allowed but not without limits and not for any purpose. There are limits to interventions, e.g., the others' well-being, and one's own safety.

That seems to be also a convergence with the most notorious advocates of enhancements, the transhumanists. "The Transhumanist agenda, which is to make such enhancement options safely available to all persons (…)".[31] Let us leave aside the parts of "enhancement" (intention) and "for all" (circumstance) for a

---

[27] N. P. G. AUSTRIACO, «Healthier than Healthy: The Moral Case for Therapeutic Enhancement», *The National Catholic Bioethics Quarterly* 17/1 (2017), in https://doi.org/10.5840/ncbq20171715 [5-7-2021], 49. The Italic is mine.

[28] *Ibid*. The Italic is mine.

[29] *Ibid.*

[30] Cf. R. D. L. C. BERNABE ET AL., «The risk–benefit task of research ethics committees: An evaluation of current approaches and the need to incorporate decision studies methods», *BMC Medical Ethics* 13/1 (2012), 6, in https://doi.org/10.1186/1472-6939-13-6 [21-1-2023], 6.

[31] Cf. N. BOSTROM, «A History of Transhumanist Thought», in M. RECTENWALD - L. CARL (edd.), *Academic Writing Across the Disciplines*, Pearson Longman, New York 2011, 12.

moment. Let us face the specific notion of safety. Transhumanists and people following other ethical frameworks different from the one used in this book might also find this proportionality reasoning useful to deliberate whether an intervention is recommended, expected to bring more benefits than risks, and sufficiently safe. People with other moral standards may agree that a person should not be put at an unjustified and irresponsible risk of danger. And when it happens, the responsible person should be held accountable for such a choice.

Considering the object-intention-circumstances framework, proportionality in the choice of means is a factor that should be explored and better developed, especially considering future interventions on the horizon. It is an element for choosing wisely among legitimate means in a specific situation. It is a matter of prudence, in the full sense of this virtue: the well-deliberated and committed choice of the best means to good ends. "Prudence is the virtue that disposes the individual not only to discern the true good in every circumstance, but also the right means of achieving it."[32] Nowadays, cowardice is often considered prudence, and courageous sacrifice is labelled imprudence. But in truth, the prudent person is not fragile nor afraid to act. Indeed, only the prudent person can be just, temperate, and brave.[33] Prudence implies serious deliberation and at the same time the brave boldness to make final decisions. The prudent person has to estimate the situation of an act and transform this cognition of reality into a concrete decision.[34]

Prudence is "the quintessence of ethical maturity"[35] and "also the quintessence of moral freedom",[36] it surpasses moralism, casuistic, and relativism. It is the fair way to face uncertainty and risk which are so common in the global governance of new technologies: "decisions which by the very nature of prudence are concerned with things concrete, contingent, and future, there cannot be that certainty which is possible in a theoretical conclusion."[37] Prudence encompasses the capacity to estimate, with a sure instinct for the future, whether a particular action will lead to the realisation of the goal. The prudent person does not expect certainty where it cannot exist and does not trust false certainties.[38] In the scientific and governance fields, prudence is what makes the stakeholders deliberate, decide, and execute a specific act well, here and now, and with concrete means, pursuing their and their community's true good.

The present search for proportionality as a guide for choosing adequate means for interventions, especially considering enhancement options on the horizon, is

---

[32] N. P. G. AUSTRIACO, *Biomedicine and Beatitude: An Introduction to Catholic Bioethics*, The Catholic University of America Press, Washington, D.C. 2021(2), 17–18.

[33] Cf. J. PIEPER, *The Four Cardinal Virtues*, University of Notre Dame Press, Notre Dame 1966, 3–5.

[34] Cf. *Ibid.*, 31.

[35] *Ibid.*

[36] *Ibid.*

[37] *Ibid.*, 18.

[38] Cf. *Ibid.*

an exercise of proactive Bioethics attempting to offer criteria for prudent deliberation. Firstly, some previous uses of proportionality were explored. For instance, proportionality between crime and punishment. In the legal field, the principle of proportionality has its origins in German case law[39] and was later incorporated in the Treaty of Amsterdam.[40] It was also proposed to deliberate how much force is appropriate in military interventions.[41]

Principles of proportionality have also appeared in medical decision-making contexts, e.g., terminal care[42] pandemics,[43] and invasive new neurotechnologies.[44] Regarding biomedical research decisions, proportionality is an idea somehow included in the *Helsinki Declaration*, in the articles concerning risks, benefits and burdens.[45] Regarding the dominant bioethical framework in Anglo-Saxon countries, inspired by Beauchamp and Childress' *Principles of Biomedical Ethics*, proportionality was only mentioned in the context of utility and double effect rule.[46]

In the field of ethics, Göran Hermerén affirms that, as an ethical principle, proportionality must be clarified and used with flexibility, as a context-dependent principle.[47] He writes: "The moral version of the principle of proportionality is about the relation between ends and means. The general idea underlying this principle seems to be that the relationship between ends and means should be 'appropriate' or 'adequate'."[48] Hermerén works with some conditions:

---

[39] Cf. M. Poto, «The Principle of Proportionality in Comparative Perspective», *German Law Journal* 8/9 (2007), 835–869.

[40] Cf. «Treaty of Amsterdam amending the Treaty on European Union, the Treaties establishing the European Communities and certain related acts», 1997, 1–144, in http://data.europa.eu/eli/treaty/ams/sign/eng [28-1-2023].

[41] Cf. J. Forge, «Proportionality, Just War Theory and Weapons Innovation», *Science and Engineering Ethics* 15/1 (2009), 25–38.

[42] Cf. T. E. Quill - B. Lo - D. W. Brock, «Palliative options of last resort: a comparison of voluntarily stopping eating and drinking, terminal sedation, physician-assisted suicide, and voluntary active euthanasia», *JAMA* 278/23 (1997), 2099–2104.

[43] Cf. J. J. Fins - F. G. Miller, «Proportionality, Pandemics, and Medical Ethics», *The American Journal of Medicine* 133/11 (2020), 1243–1244, in https://www.ncbi.nlm.nih.gov/pmc/articles/PMC7347467/ [26-4-2022], 1244.

[44] Cf. P. Kellmeyer, *Proportionality as a Guiding Principle for Regulating Invasive Neurotechnological Medical Devices*, in (2017), poster 58, in https://www.ieeg.uni-freiburg.de/data/files/pdf/kellmeyer-INS2/view?searchterm=proporti [26-4-2022].

[45] Cf. The World Medical Association (WMA), «Declaration of Helsinki – Ethical Principles for Medical Research Involving Human Subjects», 1964, in https://www.wma.net/policies-post/wma-declaration-of-helsinki-ethical-principles-for-medical-research-involving-human-subjects/ [28–1-2023] 16–18.

[46] Cf. T. L. Beauchamp - J. F. Childress, *Principles of Biomedical Ethics*, Oxford University Press, New York 2001 (5).

[47] Cf. G. Hermerén, «The principle of proportionality revisited: Interpretations and applications», *Medicine, Health Care and Philosophy* 15/4 (2012), 373–374.

[48] *Ibid.*, 374.

- Importance of the goal.
- Relevance of means to achieve the goal.
- Most favourable option (there is no other less controversial or risky means to achieve the goal).
- Non-excessiveness (of the means in relation to the goal) or Least-harmful alternative.[49]

The author also says "Means can be proportional or adequate for reaching unimportant goals if their adverse effect is low."[50] He shows how the conditions work, in which cases, and their flaws. In the final remarks, Hermerén says that proportionality is an ethical principle that might differ depending on the ethical framework. Questions can be raised, framed, and answered differently depending on the ethical context (utilitarianism, human rights, Kantianism, etc.).[51] However, he values the awareness it raises about the relationship between ends and means. The principle of proportionality "is useful in the practical analysis of pros and cons of different research strategies and treatment options. (…) In general, it may be useful in raising awareness of the problematic and context-dependent relations between end and means."[52]

Having explored previous uses of proportionality, the next chapter offers a new proposal of proportionality to help ethical deliberation in cases of interventions. Proportionality is not the only principle, but it seems the current missing factor for choosing the right means for interventions.

[49] Cf. G. Hermerén, «The principle of proportionality revisited: Interpretations and applications», *Medicine, Health Care and Philosophy* 15/4 (2012), 373–382.

[50] *Ibid.*, 375.

[51] Cf. *Ibid.*, 381.

[52] *Ibid.*

7 Principle of Proportionality for the Choice Among Legitimate Means for Intervention

Abstract

Recognising the limits of "therapy versus enhancement" as the criteria for morality, and considering the previous analysis on the object, intentions, and circumstances of enhancement, Bioethics should propose the conditions for ethical enhancements. An initial idea was developed: a framework that is not based exclusively on the intention to allow or prohibit the use of any means, does not label a means as inherently bad only because they could be misused, and recognises different necessity levels and risk levels associated with the means. The adequate selection among legitimate means for interventions implies proportionality: risky means are proportionally permitted to the level of necessity. Therefore, riskier means can be applied for higher necessities, while there is more liberty to use low-risk means. This chapter explains the principle of proportionality and its elements (necessity levels and means risk levels). It then presents examples of applications of the principle of proportionality. Lastly, some clarifications are made, showing the key strengths of this principle, and its consequences for the ethical analysis of enhancement interventions.

Keywords

Health • Therapy • Enhancement • Ethics • Proportionality

7.1 Principle of Proportionality

Recognising the limits of "therapy versus enhancement" as the criteria for morality, and considering the previous analysis on the object, intentions, and circumstances of enhancement, Bioethics should propose the conditions for ethical enhancements. An initial idea was developed: a framework that is not based exclusively on the

L. Santos, *Governance of Human Gene Editing and Transhumanism*,
Integrated Science 28, https://doi.org/10.1007/978-3-031-70863-3_7

intention to allow or prohibit the use of any means, does not label a means as inherently bad only because they could be misused, and recognises different necessity levels and risk levels associated with the means.[1]

The adequate choice among legitimate means for interventions implies proportionality in these terms:

- Risky means are proportionally permitted to the level of necessity.
  Riskier means are allowed at higher necessity levels because the benefit outweighs the risk in such cases. However, riskier means are not allowed at lower necessity levels, because the risk outweighs the benefit in such cases.
- Low-risk means are allowed at any necessity level.
  To make it even shorter for application: *riskier means—higher necessities*, and *low-risk means—liberty*.

The elements of the principle of proportionality are shown in Fig. 7.1 and an explanation of each category follows.

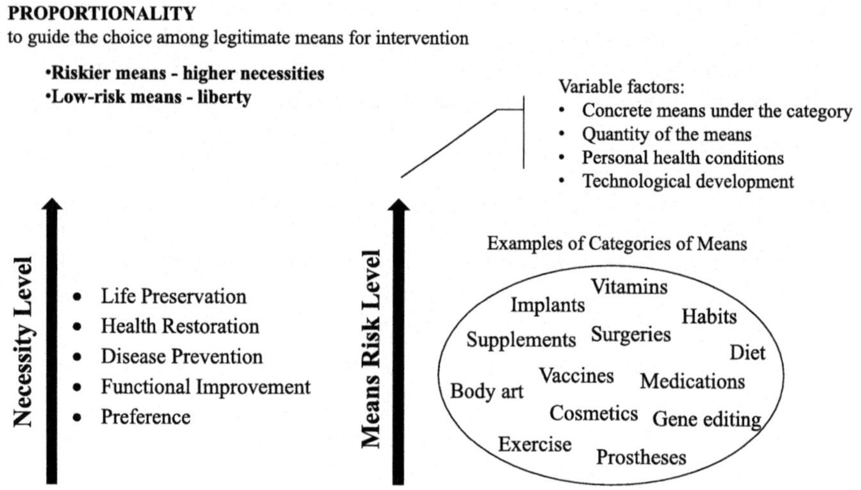

Fig. 7.1 Principle of proportionality[2]

[1] This framework is synthesised in the second part of the article: L. SANTOS, «Transhumanism and Global Governance of Human Genome Editing: Common Themes and Implications for Bioethics», *Medicina y Ética* 34/4 (2023), in https://doi.org/10.36105/mye.2023v34n4.05 [2-10-2023].

[2] L. SANTOS, «Transhumanism and Global Governance of Human Genome Editing: Common Themes and Implications for Bioethics», *Medicina y Ética* 34/4 (2023), in https://doi.org/10.36105/mye.2023v34n4.05 [2-10-2023], 1151.

7.1.1 Necessity Levels

Necessity is a requirement, meaning a condition needs to be met. For example, humans need to drink water to survive. What is considered a necessity? Are there different necessity levels? Liberal societies risk considering necessity as the object of personal wishes, hierarchising needs according to the intensity of individual desires. This approach reduces objective needs to mere wishes. It leads to relativism and renders the principles of deliberation useless, as value is given by what the individual desires. The consequences can be the domination of the strongest or, if the minimum "do not harm others" is respected, this almost unlimited liberty may harm the individual himself.

To avoid this, a list of human necessities should be found. The most known tentative to identify and order necessities is Maslow's hierarchy of needs. Psychologist Abraham Maslow analysed human motivation and individuated a set of human basic needs. In *A Theory of Human Motivation*, published in 1943, Maslow showed that there are at least five sets of human goals that may be called basic needs: physiological, safety, love, esteem, and self-actualisation. According to the psychologist, we are motivated by the desire to achieve and maintain the conditions of these basic needs.[3]

> It is quite true that man lives by bread alone — when there is no bread. But what happens to man's desires when there is plenty of bread and when his belly is chronically filled?
>
> *At once other (and "higher") needs emerge* and these, rather than physiological hungers, dominate the organism. And when these in turn are satisfied, again new (and still "higher") needs emerge and so on. This is what we mean by saying that the basic human needs are organized into a hierarchy of relative prepotency.[4]

This theory received critiques and reviews.[5] Maslow continued to refine his theory over several decades.[6] He proposed that the order of the hierarchy "is not nearly as rigid"[7] as it might have seemed at first. While it was true that most of the people seemed to have these basic needs in the order that had been indicated, there were some exceptions (e.g., some artists and martyrs).

Rather than true exceptions, these cases support his theory. Even when someone lays down his life for a higher purpose- which seems to be an inversion in the

[3] Cf. A. H. MASLOW, «A theory of human motivation», *Psychological Review* 50/4 (1943), 370–396.

[4] *Ibid.*, 375.

[5] Cf. M. A. WAHBA - L. G. BRIDWELL, «Maslow reconsidered: A review of research on the need hierarchy theory», *Organizational Behavior & Human Performance* 15 (1976), 212–240.

[6] Cf. A. H. MASLOW, *Motivation and personality*, Harper's psychological series, Harper & Row, Publishers, New York 1954; Cf. A. H. MASLOW, *Motivation and personality*, Harper & Row, New York 1970²; Cf. A. H. MASLOW - R. COX, *Motivation and personality*, edd. R. FRAGER - J. FADIMAN - C. MCREYNOLDS, Pearson Education, Inc., Uttar Pradesh, India 1987³; Cf. A. H. MASLOW, *Religions, values, and peak-experiences*, Penguin Arkana, New York 1994.

[7] A. H. MASLOW - R. COX, *Motivation and personality*, Harper & Row, New York 1970 (2), 68.

hierarchy of needs- it might be a sign that this person had the other basic levels fulfilled throughout his life, making him able of sacrifice in that specific moment. In Maslow's words: "People who have been satisfied in their basic needs throughout their lives, particularly in their earlier years, seem to develop exceptional power to withstand present or future thwarting of these needs simply because they have strong, healthy character structure as a result of basic satisfaction."[8]

Another factor worth mentioning is that the hierarchy of needs might be affected in specific situations based on individual differences and external circumstances. Maslow also points out that "any behavior tends to be determined by several or all of the basic needs simultaneously rather than by only one of them",[9] recognising that most behaviours are multi-motivated.

Another valuable aspect is that these basic needs are cross-cultural. Maslow recognised that most startling cultural differences are superficial rather than basic. For example, differences in style of hair, clothes, tastes in food, etc. His classification of basic needs is in part an attempt to account for the unity that exists behind the apparent diversity from culture to culture. Human basic needs are far more common than superficial desires. "No claim is made that it is ultimate or universal for all cultures. The claim is made only that it is relatively more ultimate, more universal, more basic, than the superficial conscious desires from culture to culture, and makes a somewhat closer approach to common-human characteristics."[10] In short, "Basic needs are more common-human than superficial desires or behaviors."[11]

Maslow's idea of *human basic needs,* and the *connection between these needs and human actions* are especially useful for the present research. There is a valuable link between objective human necessities and subjective desires and choices. In Maslow's framework, not arbitrary wishes, but basic human needs are what motivate deep desires and actions.

Other authors, from a philosophical and theological perspective, have developed similar theories about human basic needs. For instance, Aquinas wrote about connatural inclinations imprinted within the human heart. He individuated four subordinate ends that reason discovers: life, procreation, community, and truth.[12] Germain Grisez affirmed that there are human goods that contribute to the fulfilment of the individual and the community, and defended the universality of these basic goods.[13] He divided human goods into seven categories: self-integration, practical reasonableness and authenticity, justice and friendship, religion, life and health, knowledge of truth and appreciation of beauty, and satisfaction in playful

[8] A. H. MASLOW, «A theory of human motivation», *Psychological Review* 50/4 (1943), 387–388.

[9] A. H. MASLOW - R. COX, *Motivation and personality*, Harper & Row, New York 1970 (2), 71.

[10] A. H. MASLOW, «A theory of human motivation», *Psychological Review* 50/4 (1943), 389–390.

[11] *Ibid.*

[12] Cf. T. AQUINAS, «Thomas de Aquino, Opera omnia», 2019, *Corpus Thomisticum*, in https://www.corpusthomisticum.org/iopera.html [12-12-2020] I–II, q. 94, a. 2.

[13] Cf. G. GRISEZ - R. SHAW, *Beyond the New Morality: The Responsibilities of Freedom*, University of Notre Dame Press, Notre Dame 1988^3, 68–69.

activities and skilful performances.[14] John Finnis also proposed a list of basic human goods, distinguishing some equally valuable basic goods: life, health, knowledge, play, friendship, religion, and aesthetic experience.[15]

Beyond these scholars, basic human needs are related to the modern human rights framework. Human rights respond to human basic needs. It should be noted that the *Universal Declaration of Human Rights*[16] and the documents that inspired from the *International Bill of Human Rights*,[17] do not propose a hierarchy in the way that Maslow does. Rather, they stress the interconnectedness and interdependence of human rights. So, why cite a framework that does not present a hierarchy? Wouldn't this be against the very point of proposing a hierarchy of needs in the first place?

Firstly, the choice of listing these rights as "human rights" shows the recognition that certain rights are more essential than others. Therefore, there is a distinction between the basic rights stated in the *Universal Declaration of Human Rights* and the various other rights specified in other codes. The same can be said for the theories that call some human needs "*basic* human needs".

Secondly, although the Declaration did not create a hierarchy and rather stressed the interdependence of human rights, it is not incompatible with attempts to make a hierarchy among these fundamental rights. For instance, in a publication celebrating the 20th anniversary of the UNESCO *Universal Declaration of Bioethics and Human Rights*,[18] former UNESCO Director-General, Mayor Zaragoza, affirms:

---

[14] Cf. G. G. Grisez, *Christian Moral Principles: Way of the Lord Jesus*, I, Franciscan Pr, Chicago 1983.

[15] Cf. J. Finnis, *Natural Law and Natural Rights*, OUP Oxford, Oxford 2011.

[16] Cf. UN, «Universal Declaration of Human Rights», 1948, in https://www.ohchr.org/en/human-rights/universal-declaration/translations/english [13-7-2019].

[17] In 1948, for the first time, countries agreed on a list of inalienable human rights, the *Universal Declaration of Human Rights* (UDHR). Although the Declaration is "soft law" and not legally binding, it has an immense ethical and political value. In 1966, the UN General Assembly adopted two international treaties inspired on the UDHR: the *International Covenant on Economic Social and Cultural Rights* (ICESCR), and the *International Covenant on Civil and Political Rights* (ICCPR). These last two are international "hard law" and therefore legally binding. Together, the UDHR, the ICESCR, ICCPR are known as the *International Bill of Human Rights.* Cf. UN, «International Bill of Human Rights», *OHCHR*, in https://www.ohchr.org/en/what-are-human-rights/international-bill-human-rights [6-5-2022].

For more information about the main international treaties concerning human rights, Cf. UN, *The core international human rights treaties*, United Nations 2014, in https://doi.org/10.18356/9d44ec81-en [22-3-2023].

[18] Cf. UNESCO, «Universal Declaration on Bioethics and Human Rights» in http://portal.unesco.org/en/ev.php-URL_ID=31058&URL_DO=DO_TOPIC&URL_SECTION=201.html [8-6-2019], 80.

"The basic ethical reference is Human Rights which are indivisible, inherent, universal and inalienable. But there is one right, the right to life, which takes priority, because the exercise of all these rights presupposes life."[19]

Lastly, the human rights framework supports the idea that these fundamental rights respond to basic human needs. All humans have these basic needs, and no individual can ever lose or give away their human rights. These rights are the conditions for human flourishing. These fundamental rights relate to the most important needs. Human rights belong to every human being, they are not a favour the State grants. Human rights exist because we are humans, not because the law "creates" them. Even though our understanding of human rights and duties develops over time, humans always have basic needs and therefore basic rights. Afterwards, the law is called to protect these basic human rights. That is why, if a law goes against a human right or a basic need, it is defined as an unjust law. Regarding important matters where laws violate a human right or a basic need, a conscientious objection is allowed and even morally required.

Having explained the above, it can be noted that Maslow's hierarchy of needs, as well as the human rights framework, both propose levels of necessity regarding modifications. Table 7.1 compares the necessity levels for interventions proposed in this book, Maslow's hierarchy of needs,[20] and the principles of the *Universal Declaration of Human Rights*.[21]

Therefore, the necessity levels for intervention proposed in this work are as follows (from highest to lowest):

- *Life preservation*: actions aiming to save life (e.g., resuscitation).
- *Health Restoration*: actions aiming to achieve normal levels of health (e.g., hormonal treatment).
- *Disease Prevention*: actions aiming to avoid illness (e.g., vaccines).
- *Functional Improvements*: actions aiming to increase the performance of a useful trait (e.g., increasing concentration).
- *Preferences*: actions that produce no objective improvement, based only on personal desires and preferences (e.g., changing hair colour).

As shown in Table 7.1, *life preservation* and *health restoration*- are related to the *physiological needs* seen in Maslow's classification. They have to do with the biological requirements for human survival. *Disease prevention* is related to *security needs*, since the person is healthy and the requirements for survival are met, but

[19] F. MAYOR ZARAGOZA, «Dignity as the foundation for all Human Rights», in G. SOLINÍS (ed.), *Global bioethics: what for? Twentieth anniversary of UNESCO's Bioethics Programme*, UNESCO, Paris 2015, 47–51, 48.

[20] Cf. A. H. MASLOW, «A theory of human motivation», *Psychological Review* 50/4 (1943), 370–396.

[21] Cf. UN, «Universal Declaration of Human Rights», 1948, in https://www.ohchr.org/en/human-rights/universal-declaration/translations/english [13-7-2019].

Table 7.1 Comparative chart of human needs

Necessity level for interventions	Maslow's hierarchy of needs	Universal declaration of human rights
Life preservation and health restoration	*Physiological needs* Biological requirements for human survival (e.g., air, food, drink, shelter, clothing, warmth, sleep)	3. Life, liberty and security 5. No torture, cruel, inhuman or degrading treatment or punishment 25. Food, clothing, housing and medical care and necessary social services, social security 24. Rest and holiday
Diseases prevention	*Safety needs* Protection (e.g., security, order, law, stability, freedom from fear)	4. No slavery 6. Legal personality 7. Equal protection before the law 8. Fair treatment by court 9. No unfair detainment 10. Fair trial 11. Innocent until proved guilty 13. Movement 14. Seek asylum from persecution 17. Property 30. No destructive interpretation of rights
Functional improvements and preferences	*Love needs* Belongingness (e.g., friendship, intimacy, trust, and acceptance, receiving and giving affection, and love)	15. Nationality 16 Marriage and family 20. Peaceful assembly and association 21. Political participation 28. Social and international order 29. Duties to the community
	Esteem needs Self-worth and respect (e.g., dignity, achievement, independence, reputation, status, prestige)	1. Born free and equal 2. No discrimination 12 Privacy, honour and reputation 23. Labour rights
	Self-actualization needs Realization of a person's potential (e.g., self-fulfilment, seeking personal growth and peak experiences, creativity)	18. Freedom of thought, conscience and religion 19. Freedom of opinion and expression 22. Economic, social and cultural rights 26. Education 27. Cultural life, arts, and science

the person is safer if diseases are prevented. Lastly, *functional improvements* and *preferences* are related to human needs for *love, esteem,* and *self-actualization.*

Maslow noted that his order is not rigid[22] and that some behaviours are multi-motivated.[23] This can also be seen in the case of interventions. For instance, some surgeries treat a deviated septum and they turn into a nose reshaping procedure at the same time. Furthermore, some prostheses restore functioning and add on other features, either aimed at functional improvement or individual preference. For example, James Young lost his arm and leg in an accident in 2012. His limbs were replaced by robotic prostheses, and controlled by sensors that convert nerve impulses from Young's spine into physical movements. But the prosthesis went beyond the arm's functions featuring a USB port, a screen displaying his Twitter feed and a retractable dock containing a remote-controlled drone.[24] This example relates to the *restoration of health*, but also *functional improvements* and *personal preferences.*

While Young's robotic prostheses did not mean a risk to his health, there are cases in which some people risk their health and even their lives to achieve a functional improvement or a personal preference. For example, there is nothing wrong with aiming to be thin. But if a person with an average weight aims to be thinner and thinner to the point of compromising health, this person will suffer from anorexia or some similar *dis*-order. The word *disorder* shows the idea that there is in fact an order, a hierarchy. One could go on analysing what lies behind the desire to be thin, and consider changes in beauty standards, lack of acceptance and need for self-esteem, or the dangers of social pressures. However, in itself, to be thin under certain parameters is something neutral.

Similarly, there is nothing wrong with making changes according to personal preference. We are both biological and cultural beings. Our ideas and free choices shape some bodily aspects. However, if life or health is seriously endangered by choices based on personal preference, this might signal a disorder.

Concluding this section regarding necessity levels, there is a list of needs that are more important than others, a set of basic human goods. Considering the topic addressed in this thesis, the proposed order is as follows: first, *life preservation*, then *health restoration*, followed by *disease prevention*, then *functional improvements*, and lastly *personal preferences.*

---

[22] Cf. A. H. MASLOW - R. COX, *Motivation and personality*, in edd. R. FRAGER - J. FADIMAN - C. MCREYNOLDS, Pearson Education, Inc., Uttar Pradesh, India 1987 (3), 68.

[23] Cf. *Ibid.*, 71.

[24] Cf. K. TSUI, «Transhumanism: Meet the cyborgs and biohackers redefining beauty», *CNN* (2020), in https://www.cnn.com/style/article/david-vintiner-transhumanism/index.html [22-3-2021].

7.1.2 Means Risk Levels

Means refers to the method or tool implied to intervene or to modify a given reality. Intervention in this work refers to an action that changes the current state of the human body. Presented below are some categories of means and their examples:

- *Gene editing*: tools that alter an organism's DNA (e.g., CRISPR-Cas9).
- *Implants*: insertion of a device inside or on the surface of the body (e.g., pacemakers).
- *Prostheses*: replacements for body parts that are missing or malfunctioning (e.g., artificial leg).
- *Surgeries*: cutting into the skin to access internal parts of the body (e.g., caesarean section).
- *Medications*: chemical substances that produce a biological effect (e.g., antibiotics).
- *Vaccines*: substances used to stimulate immune response against diseases (e.g., polio vaccine).
- *Vitamins and Supplements:* addition of nutritive value (e.g., vitamin C).
- *Habits, diet, exercise*: behaviours (e.g., low cholesterol diet).
- *Cosmetics*: substances used to change complexion or appearance (e.g., hair colours).
- *Body art:* body modifications with artistic meaning (e.g. tattoos).

This list is not meant to be exhaustive. In the future other types of means will emerge. Also, this list is not meant to be hierarchical. There is a general risk level associated with each category, but this is much less stable than the necessity levels. Risk classification depends on *variable factors* such as:

- *The specific means under the category.* The general categories presented here portray that the risk level is not the same regarding all types of interventions. But inside each category, there are multiple types of interventions, and therefore multiple associated risk levels. For instance, under the category of gene editing, there are many types of genetic interventions, each one possessing a different risk level. For example: in vivo is riskier than ex vivo, heritable HGE is riskier than somatic HGE, some genes are riskier to edit than others, etc. Furthermore, some medications are less risky than some implants (e.g., paracetamol versus a pacemaker), and some implants are less risky than some medications (e.g., a dental implant versus chemotherapy).
- *The quantity of the means.* Although an intervention has a specific initial risk, if the means is repeatedly used or used in high amounts, its risk increases. For example, a caesarean surgery has a level of risk, but a fourth caesarean is considerably riskier. An example regarding a low-risk means is that coffee has a low-risk level, but coffee in large amounts and over time causes addiction and health problems.

- *Personal health conditions.* Some options are riskier for one person than they are for another. If someone has a contraindication regarding a specific medication, a type of gene editing could be a better option. Yet for a person who has no contraindication to the same given medication, the drug would be less dangerous than a genetic intervention.
- *Technological development.* The associated risk level depends on current technological advancements. For example, current surgeries are more precise and less risky than the surgical options available in the past. One can reflect on how surgeries were before and after anaesthesia, how medicines were before and after antibiotics, how personalised medicine was before and after gene editing, wearable devices, AI, etc. When applying the principle of proportionality, the current stage of technological development must be considered. But without forgetting that, in time, other means might be created, or an existing means may become less risky. This would add options and change the hierarchy. For example, gene editing might become more efficient with time, and serve as an alternative to other aggressive treatments such as chemotherapy.

To conclude this section regarding the means risk levels, even if the categories of means proposed as examples change depending on variable factors, the reasoning behind the principle of proportionality remains valuable as a guideline. Risky means are allowed in proportion to the level of necessity. Riskier means are justified for higher necessities, while there is more liberty in the use of low-risk means.

7.2 Examples of Application

I hope to make clearer the usefulness of the proportionality principle by applying it to a couple of examples, some real and some hypothetical.

In synthesis: the same means risk level (transplant), and a different necessity level (health restoration versus functional improvement). A heart transplant is a risky intervention, but it is morally acceptable for the restoration of health. However, it should not be done for a functional improvement. An athlete with a healthy heart should not take the risks of a heart transplant with the aim of gaining a stronger heart.

Table 7.2 Heart transplant

Heart transplant		
Necessity	Means	Principle applied
Health restoration (heart failure)	Heart transplant	YES Riskier means—higher necessities
Functional improvement (stronger heart)		NO Riskier means—higher necessities

If an unjustified heart transplant is ever done on a healthy person to increase their athletic ability, the unfortunate case would not make all heart transplants unethical. It would make this specific transplant unethical, as the means is disproportional to the necessity.

Table 7.2 shows the same kind of intervention (heart transplant), to the same general risk level. Clearly, variable factors such as health conditions cause risks to increase.

It should not be forgotten that the principle of proportionality serves the choice among legitimate means. However, there are other factors to be considered in the bioethical analysis, such as the informed consent of the donor or of relatives, fairness (waiting list of people depending on a donated heart to keep living), etc.

Let us give another similar example, with a less common means.

In synthesis: the same means risk level (implant), and a different necessity level (health restoration versus preference). Following the principle of proportionality, we should not endanger someone by inserting a chip into their brain to control their body movements, on the basis-the of personal desire. But inserting a chip in the brain of a person to restore his autonomy regarding bodily movement would be a different case. Hopefully, that is not so far in the future. In 2023, a paralysed man has been able to walk thanks to electronic brain implants.[25] In the same year, a study on a spinal cord neuroprosthesis for locomotor deficits due to Parkinson's disease was published.[26]

In short, Table 7.3 presents the same kind of modification and the same risks of side effects. What is the difference? The different levels of necessity, justifying or not the application of risky means. In contrast, Table 7.4 follows an example in which there is the same necessity level but different options of means, implying different risk levels.

Table 7.3 Chip in the brain

Chip in the brain to control movements		
Necessity	Means	Principle applied
Health restoration (loss of autonomous movements)	Implant (chip in the brain to control movements)	YES Riskier means—higher necessities
Preference		NO Riskier means—higher necessities

[25] Cf. P. GHOSH, «Brain implants help paralysed man to walk again», *BBC News* (2023), in https://www.bbc.com/news/science-environment-65689580 [27-5-2023].

[26] Cf. T. MILEKOVIC ET AL., «A spinal cord neuroprosthesis for locomotor deficits due to Parkinson's disease», *Nature Medicine* 29/11 (2023), 2854–2865, in https://www.nature.com/articles/s41591-023-02584-1 [24-5-2024].

Table 7.4 Interventions to improve concentration in a person with an *average* capacity level

Interventions to improve concentration in a person with an *average* capacity level		
Necessity	Means	Principle applied
Functional improvement (increasing average concentration capacity)	Gene editing(hypothetical somatic HGE)	NO Riskier means—higher necessities
	Medication (Ritalin)	NO Riskier means—higher necessities
	Diet (coffee)	YES Less risky means—liberty

In synthesis: the same necessity level (functional improvement), and different means risk level (gene editing, medication, diet). In the previous examples, Table 7.2 (heart transplant) and Table 7.3 (chip in the brain), the intervention was the same while the necessity level varied. In this example, the necessity level is the same (improvement), but the risk level changes according to the proposed means (gene editing, medicine, diet).

Drinking coffee differs from taking drugs or using gene editing to improve concentration. According to the principle of proportionality, there is more freedom to choose among low-risk means. However, it should be remembered that the variable factor of the quantity of the means because even low-risk means can cause harm if used without balance.

What about a scenario with the same means and same risk level, but instead of a functional improvement as the necessity level, health restoration? In other words, instead of someone with an *average* capacity level, someone with a *below-average* concentration level. That is the scenario presented in Table 7.5.

In synthesis: the same necessity level (health restoration), and different means risk level (diet, medication, gene editing). In this situation, the proposed modifications are the same as in the previous Table 7.4 (diet, medicines, gene editing), and the effect is the same as well (increasing concentration capacity).

Table 7.5 Interventions to improve concentration in a person with a *below-average* capacity level

Interventions to improve concentration in a person with a *below-average* capacity level		
Necessity	Means	Principle applied
Health restoration (achieving average concentration capacity)	Diet (coffee)	YES Less risky means—liberty
	Medication (Ritalin)	YES Riskier means—higher necessities
	Gene editing(hypothetical somatic HGE)	YES Riskier means—higher necessities

The change in Table 7.5 is the necessity level (health restoration instead of the previously considered improvement). This example is different because the person has a below-average concentration capacity. Therefore, this example is not the improvement of a normal function, but the restoration of health. As seen in the scenario of Table 7.4, low-risk modifications such as changes in diet are permitted, but in the case of higher necessity, higher levels of modification are permitted as well (in Table 7.5, medicines and hypothetical gene editing).

The variable factors must also be considered:

- *Specific modification under the category*: What are the different medicines and treatments available? Which type of HGE is being considered?
- *Quantity of the modification*: Does the amount of medication or HGE modifications change the risk consideration?
- *Personal health conditions*: Is this a serious concentration problem? What causes the lack of concentration in this person? What are the health conditions of this person? What are the potential risks and benefits for this specific person?
- *Technological progress*: What is the treatment available now? Is there a type of HGE capable of solving this problem? Is this medicine/this gene editing more efficient than other alternatives? Is it more dangerous than other alternatives? For instance, today Ritalin is less risky than an experimental attempt of HGE to improve concentration. If later in time gene editing has fewer adverse effects than this specific drug, the conclusion should accompany this technological advancement.

Now, let us apply the principle of proportionality to interventions to achieve average cholesterol levels. LDL (low-density lipoproteins) are known as "bad" cholesterol because a high LDL level leads to a buildup of cholesterol in the arteries.

In synthesis: Table 7.6 presents the same necessity level (health restoration), and different means risk level (gene editing, medicines, diet). It shows interventions to achieve below-average cholesterol levels, which is better for an individual's health.

Table 7.6 Interventions to achieve *below-average* cholesterol levels

Modifications to achieve *below-average* cholesterol levels		
Necessity	Means	Principle applied
Disease prevention(achieving better cholesterol levels)	Diet (low cholesterol diet)	YES Less risky means—liberty
	Medicines (Statins)	YES Riskier means—higher necessities
	Gene editing(somatic HGE)	YES Riskier means—higher necessities

We could also hypothesise different necessity levels (health restoration on the one hand, and disease prevention on the other hand). Both intentions are good. Health restoration and disease prevention could be considered under the traditional therapeutic goal of medicine. However, this is one of the cases in which enhancement can have a therapeutic use, preventing diseases.

Nicanor Austriaco has suggested that lowering LDL levels to 50 mg/dL can be justified alongside other lifestyle modifications to lower cholesterol levels, since the studies published so far suggest that the benefits of such modifications vastly outweigh any of its burdens.[27] Following Austriaco's suggestion, Table 7.6 hypothesises that using HGE to lower cholesterol levels would be better for health and prevent cardiovascular diseases. Still, this intervention has yet to pass current clinical trials.[28] HGE to reduce the risk of cardiovascular disease "could only be justified if the benefits of such an intervention outweigh any accompanying burdens. Persons can only be made healthier than healthy if they do not unreasonably put their health at risk."[29]

Other authors also call attention to the difference between gene editing and other less risky medical practices concerned with the same objective of lowering LDL levels: "There are notable dissimilarities between genetic interventions and less invasive medical practices with the same objective; e.g. there is a vast distinction between genome editing and taking medication to reduce cholesterol levels."[30] Mendz and Cook affirm that "Regardless of the similarity of the intent, the actual nature of the action makes them significantly different."[31]

At the moment, there are less risky medical practices other than HGE, which is still in clinical trials, that can be used to achieve better LDL levels. Although the goal and outcome might be the same, less risky means should be selected for the good of the individual. However, variable factors such as technological progress should not be forgotten. Perhaps, in the future, gene editing will develop to a point where it will have a much lower risk than today, and maybe it will have fewer adverse effects than some medicines. If this happens, following the same principle of proportionality, reflections could change, favouring gene editing over medications like Statins.

[27] Cf. N. P. G. Austriaco, «Healthier than Healthy: The Moral Case for Therapeutic Enhancement», *The National Catholic Bioethics Quarterly* 17/1 (2017), in https://doi.org/10.5840/ncbq20171715 [5-7-2021], 49.

[28] Cf. Verve Therapeutics, «Verve Therapeutics Doses First Human with an Investigational In Vivo Base Editing Medicine», VERVE-101, 2022, *Verve Therapeutics*, in https://ir.vervetx.com/news-releases/news-release-details/verve-therapeutics-doses-first-human-investigational-vivo-base/ [3-2-2023].

[29] N. P. G. Austriaco, «Healthier than Healthy: The Moral Case for Therapeutic Enhancement», *The National Catholic Bioethics Quarterly* 17/1 (2017), in https://doi.org/10.5840/ncbq20171715 [5-7-2021], 49.

[30] G. Mendz - M. Cook, «Transhumanist Genetic Enhancement: Creation of a 'New Man' Through Technological Innovation», *The New Bioethics* 27/2 (2021), 116. Underline is mine.

[31] *Ibid.* Underline is mine.

Table 7.7 Interventions to improve musical ability in the unborn

Interventions to improve musical ability in the unborn		
Necessity	Means	Principle applied
Functional improvements (improving musical ability)	Habits (playing Mozart)	YES Less risky means—liberty
	Gene editing(somatic in utero HGE)	NO Riskier means—higher necessities
	Gene editing (heritable HGE)	NO Riskier means—higher necessities

Lastly, this next example explores interventions in the unborn, Table 7.7 includes a differentiation between somatic and heritable gene editing in terms of proportionality.

This example echoes Bostrom's comparison that was previously analysed in this work. The transhumanist complained that some people would accept playing Mozart to an unborn child, but reject genetic interventions to reach the same goal. The quote presented below is worth noting:

> Nobody would argue for a ban on Mozart-in-the-womb on grounds that we cannot rule out that some psychological woe might befall the child once she discovers that her facility with the violin had been prenatally 'programmed' by her parents. Yet when, for example, it comes to genetic enhancements, eminent bioconservative writers often put forward arguments that are not so very different from this parody as weighty, if not conclusive, objections. To transhumanists, this looks like doublethink.[32]

Bostrom did not consider that playing music and genetic interventions are very different means to pursue the same outcome. The controversies are not due to the improvement *intention*, but rather to the different *means* proposed. The improvement of musical talent is not bad, but it should be pursued through less dangerous means.

Furthermore, the problem is not intrinsically related to the mean (gene editing), but to the *lack of adequacy* of its use in this specific case. Mozart-in-the-womb may have benefits for the unborn child while bringing a low or inexistent risk for the pregnant woman and the child. In contrast, both *in uterus* somatic HGE or heritable HGE would put the mother and child in a risky situation, which is not proportionally justified to an increase in the child's musical ability.

In uterus somatic HGE is not a bad means in itself, as it has the potential to be a type of precise surgery for the not-yet-born. Although *in uterus* somatic HGE is disproportional and inadequate to try to increase musical talent, it should not be

[32] N. Bostrom, «In Defense of Posthuman Dignity», *Bioethics* 19/3 (2005), in https://onlinelibrary.wiley.com/doi/abs/10.1111/j.1467-8519.2005.00437.x [10-4-2021], 212.

discarded in all cases. Perhaps *in uterus* somatic HGE might be an adequate means to treat a genetic disorder at a very early stage. We can hope that it will become a safer and more available option in the future.

Regarding heritable human gene editing (HHGE), the risks for the specific child and future generations are so great at the current stage that it is hard to hypothesise a situation in which the necessity level would justify its use.

In the Third International Summit on Human Genome Editing, held in London, in 2023, a whole section was dedicated to the question "Are there compelling reasons for heritable human genome editing?"[33] What if safe HHGE is achieved?

At the Summit, Ephrat Levy-Lahad, from Shaare Zedek Medical Center, argued that enabling biological parenthood might be a compelling reason to proceed, not because it is a better kind of parenthood, but because it is important to some people. It should be noted that this argument responds to the necessity level of preferences. The speaker continued by saying that HHGE would also avoid discarding embryos in IVF procedures.[34]

In contrast, Tina Rulli, from the University of California, defended at the same summit that reproductive HHGE does not cure disease. Curing a disease is different from creating an embryo that needs treatment. Also, considering rare genetic diseases as well as limited public resources, Rulli questions whether HHGE is socially valuable and what a proportional investment in HHGE would look like. Rulli also cited common concerns such as safety, effects on future generations, fair access, the risk of a slippery slope to designer babies and new eugenics.[35]

In my opinion, heritable HGE is a concept that should be further explored making important differentiations. Heritable HGE refers not only to editing embryos but also editing gametes or their precursor cells. The modifications made in male and female gametes and primordial germ cells are theoretically heritable too. Final ethical conclusions cannot be reached without properly considering the different types of heritable HGE. Chapter 10 of this book will explore that.

Regarding the example portrayed in Table 7.7, all kinds of somatic and heritable gene editing are considered as disproportionate means to increase the chance of musical talent in a child (functional improvement).

---

[33] Cf. «Third International Summit on Human Genome Editing», 2023, *The Royal Society*, in https://royalsociety.org/science-events-and-lectures/2023/03/2023-human-genome-editing-summit/ [7-3-2023]; Cf. The Organising Committee of the Third International Summit on Human Genome Editing, «Statement from the Organising Committee», 2023, in https://royalsociety.org/news/2023/03/statement-third-international-summit-human-genome-editing/ [7-3-2023].

[34] Cf. «Third International Summit on Human Genome Editing», 2023, *The Royal Society*, in https://royalsociety.org/science-events-and-lectures/2023/03/2023-human-genome-editing-summit/ [7-3-2023].

[35] A better explanation of Rulli's argument can be consulted in the following article. Cf. T. Rulli, «Reproductive CRISPR does not cure disease», *Bioethics* 33/9 (2019), 1072–1082, in https://onlinelibrary.wiley.com/doi/abs/10.1111/bioe.12663 [8-3-2023].

7.3 Final Remarks

7.3.1 Not the Only Factor for Bioethical Deliberation

The principle of proportionality aims to help choose among legitimate means. Proportionality is a necessary step to determine whether an intervention is worth considering in a broader ethical analysis. Regarding the ethical analysis of the object, intentions, and circumstances of an intervention, the adequacy of means (part of the analysis of the object) seems to be the most complex issue that is yet to be addressed by Bioethics.

However, the principle of proportionality does not intend to be the only determinant regarding the ethical analysis of interventions. For instance, there are proximate and remote *intentions* and *circumstances* such as equality of access, respect for diversity, the idea of perfection, external influences on desires and choices, etc. Even when a risky legitimate *means* is judged to be proportional for the case, other elements must be considered when assessing the ethics of a given intervention.

Means that are risky today may imply less risk in the future. If so, the range of low-risk may be broader. In this scenario, possibilities would increase. Personal desires would have more space, as there would be more legitimate and low-risk means to pursue them. We should pay careful attention to what we desire (e.g.: psychological dangers to people desiring a kind of body influenced by social media, even if pursued with low-risk means).

7.3.2 Key Strengths of the Principle

- It is not based exclusively on the intention to allow/prohibit the use of any means.
- It recognises different necessity and means risk levels.
- It does not label a means as inherently bad only because it could be misused or disproportional in some occasions.
- It does not exclude a legitimate risky means if there is a proportional need (a high necessity) for its use.
- It recognises that we have more freedom using low-risk means, which can be considered to fulfil any necessity level, including functional improvements and personal preferences.
- It acknowledges that the risk level of means varies depending at least on these factors: *the concrete means under the category, the quantity of the means, personal health conditions,* and *technological development.*
- It shows that interventions aiming at goals other than *life preservation, health restoration* and *disease prevention*, such as *functional improvement* or *personal preferences*, are also permitted, but with a proportional approach to avoid harm.
- It is open to possible enhancements (human intervention to improve the average trait), whether they are for health or not, if the intention and circumstances

are good, and if the means are adequate (proportional) for the given situation. This principle does not uphold any kind of enhancement only based on the improvement intention or claiming respect for individual choices. However, it does not condemn all possible interventions aiming at enhancement, as if it is wrong to try to improve. Although rejecting all enhancements would avoid the related risks, that is not a solid reasoning.

New technologies and new mentalities urge new critical analyses and proposals. In every situation, there is an opportunity to take a step towards what is ethically good. No context is so dark that nothing can be done. In the present scenario, bioethicists should look for good criteria for modifications, while recognising the increasing availability of new technologies and the emerging idea of enhancement.

Considering that improvement is surrounded by risks but is not intrinsically evil, bioethicists would better serve society by differentiating types of enhancements, pondering the intentions, considering the means employed, working to improve unwanted circumstances, and proposing guidelines to differentiate ethical from unethical enhancements.

7.3.3 Adequate Enhancements Using Frameworks Other Than Transhumanism

This chapter has proposed using the principle of proportionality to determine the adequate choice of means. This framework can help many different and well-intended people: people who want to do good and avoid evil, and who aim for moral good in the object, ends and circumstances of their actions.

It is worth noting that there is a tendency to contrast transhumanist proposals and their risks either by emphasising *circumstantial* concerns or by fighting against the *intention* of improvement. However, as explained in this work, it seems that the most accurate reasoning against some enhancements proposed by transhumanists has to do with the disproportionate use of means.

Similar to transhumanist thought, this framework does not go against human enhancement, in the sense of human intervention in the average trait to improve its performance. It also does not condemn all possible modifications to the natural state. In fact, this framework is even open to actions that do not result in any objective improvement of human performance, that is, to actions that respond to personal preference. However, this framework is certainly against enhancements that use disproportional means, which endanger people trying to improve themselves or shape themselves according to personal preference.

Using the ethical criteria that an act must be good in the object (comprising adequacy of means), intention, and circumstances, it can be concluded that enhancements that meet the ethical requirements are good. That conclusion does not imply agreement with all of the ideas proposed by Transhumanism, nor does it embrace its worldview.

Enhancement should not be identified exclusively with Transhumanism. First, because the truth is that other people can defend certain human improvements based on other currents of thought. Second, because otherwise, our societies might accept many kinds of morally good or neutral enhancements in the ways proposed by transhumanists. It is possible to defend some enhancements, while abiding by other values and ethical frameworks than the transhumanist ones. More people should work on the conditions for an ethically sound enhancement. There should be pro-enhancement thinkers other than transhumanists.

Bioethics' Role in the Global Governance of Human Genome Editing

8

Abstract

A common theme that resulted from the Reflexive Thematic Analysis (RTA) method applied to selected WHO and Humanity + texts is *global governance to minimise the risks and maximise the benefits of human genome editing*. This theme comprises the main governance elements (research, funding, public debate, policies) and the predominant values commonly found in the selected texts (autonomy, protection, and equality-solidarity). This chapter begins by considering that Bioethics is searching for its role in the global governance of human genome editing (HGE). Ethics should go beyond procedures, and do more than call for transparent, inclusive, and accountable governance. Secondly, as ethics has to do with moral agents, this chapter considers how Bioethics can contribute to the key stakeholders of the global governance of HGE: researchers, policymakers, funders, and the public. Thirdly, the values in the HGE debate are changing. Past, current and future possible values are considered. Lastly, ideas on whether, why, and how to participate in the public dialogue of gene editing are offered.

Keywords

Governance • World Health Organization • Bioethics • Human genome editing • Public debate

L. Santos, *Governance of Human Gene Editing and Transhumanism*, Integrated Science 28, https://doi.org/10.1007/978-3-031-70863-3_8

8.1 Ethics Beyond Procedures

A common theme that resulted from the RTA method applied to selected WHO and Humanity+ texts is *global governance to minimise the risks and maximise the benefits of human genome editing.*[1] As explained earlier (see Chap. 1), global governance is a complex process to manage global issues. It implies multiple agents and means. To get an idea of how governance goes beyond regulations and judicial decisions, a recent example of governance is portrayed in what happened in the U.S. after the sentence *Dobbs v Jackson*[2] in 2022. The Supreme Judges recognised that abortion presents a profound moral question. They declared that there is no basis for a Constitutional right to abortion, and that the Constitution does not prohibit the citizens of each state from regulating or prohibiting abortion. *Dobbs v Jackson* overruled earlier sentences (*Roe v. Wade*[3] and *Planned Parenthood of Pennsylvania v. Casey*[4]). However, some pharmaceutical industries, employers, businesses, and even the President rapidly offered facilities to support pregnant women seeking an abortion. The offered means offered varied from facilitating travels to places where abortion is permitted, easing access to abortive pills,[5] and erasing research data that might be related to searching for a termination of pregnancy.[6] Abortion pills via telemedicine were considered among the *10 Breakthrough Technologies 2023* by MIT Technology Review.[7] Even if this example portrays to a governance mechanism at the national level, it helps us understand how governance is far broader than creating regulations, and is involved in various aspects such as the enforcement of laws, public debate, activism, the private sector, business, funding and many others. These factors all have a role in influencing governance.

At the international level, global governance has moved away from the use of *hard law,* such as treaties, as a primary instrument. That is due to the time, resources, and efforts to negotiate, implement and enforce international law

---

[1] The full Reflexive Thematic Analysis (RTA) can be consulted in the published doctoral dissertation excerpt L. Santos, Common Themes Between Humanity + Original Documents on Transhumanism and WHO Documents on the Global Governance of Human Genome Editing, Ateneo Pontificio Regina Apostolorum, Rome 2023.

[2] Cf. «Dobbs v. Jackson Women's Health Organization 597 U.S.», 2022, in https://supreme.justia.com/cases/federal/us/597/19-1392/ [1-2-2023], 2228.

[3] Cf. «Roe v. Wade 410 U.S. 113», 1973, in https://supreme.justia.com/cases/federal/us/410/113/ [6–2-2023].

[4] Cf. «Planned Parenthood of Southeastern Pa. v. Casey 505 U.S. 833», 1992, in https://supreme.justia.com/cases/federal/us/505/833/ [6-2-2023].

[5] Cf. R. Ackermannarchive, «Abortion pills via telemedicine: 10 Breakthrough Technologies 2023», 2023, *MIT Technology Review,* in https://www.technologyreview.com/2023/01/09/1064871/abortion-pills-telemedicine-10-breakthrough-technologies-2023/ [3-2-2023].

[6] Cf. «Roe v Wade 410 U.S. 113», 1973, in https://supreme.justia.com/cases/federal/us/410/113/ [6–2-2023].

[7] Cf. R. Ackermannarchive, «Abortion pills via telemedicine»: 10 Breakthrough Technologies 2023», 2023, *MIT Technology Review*, in https://www.technologyreview.com/2023/01/09/1064871/abortion-pills-telemedicine-10-breakthrough-technologies-2023/ [3-2-2023].

demands. A variety of other instruments included in *soft law*, such as principles, codes of conduct, standards, guidelines, public–private partnerships, and certification requirements have instead been used. The new model to oversee global governance often includes the participation of other entities in addition to governments, such as industry organisations, funders, civil society groups, think tanks, international organisations, insurers, and journals.[8]

There has also been a shift regarding the ultimate goal of global governance. Instead of the objective of harmonisation of identical requirements, less formal goals have been adopted: synchronisation, alignment, coordination, and cooperation have been considered more feasible and useful. These softer goals intend to recognise and respect the expression of national cultural, historical, political, and economic differences while agreeing on common core principles and procedures.[9]

Adding to the complexity of current global governance, gene editing appears to be "turning democracy upside-down",[10] forcing its global governance to become more articulated.

> In a 'CRISPR democracy', citizens take the place of scientists when they, as biohackers, start tinkering with DNA, under the guise of 'democratizing' the life sciences. The other way around, scientists take the place of citizens when their voice becomes decisive in the governance of highly controversial technologies that may have a lasting impact on no less than the future of humankind.[11]

Different means and tools constitute the structure of governance. However, the content of governance is determined by its various moral agents. Which application of HGE will be promoted by these moral agents? Which application will be discouraged? How can Bioethics help the diverse agents of the global governance of HGE?

In the 2000s, Edmund Pellegrino wrote about contemporary Bioethics' inability to generate moral guidance, stating that there is a tendency to gradually abandon the idea of normative ethics and moral truth of any kind, and the tendency to emphasise procedures and public policy.[12] In governance, Bioethics risks focusing

[8] For further readings on global governance and the different approaches in global governance, Cf. M. Hewson - T. J. Sinclair (edd.), *Approaches to Global Governance Theory*, SUNY Press, New York 1999. Specifically on the relationship between international organizations and global governance, the following reading is recommended Cf. T. G. Weiss - R. Wilkinson (edd.), *International Organization and Global Governance*, Routledge, New York 2018.[2]

[9] Cf. G. E. Marchant, «Global Governance of Human Genome Editing: What are the Rules?», *Annual Review of Genomics and Human Genetics* 22/1 (2021), in https://www.annualreviews.org/doi/10.1146/annurev-genom-111320-091930 [16-9-2021], 393–394.

[10] J. B. Hurlbut - K. Saha - S. Jasanoff, «CRISPR Democracy: Gene Editing and the Need for Inclusive Deliberation», *Issues in Science and Technology* XXXII/1 (2015), in https://issues.org/crispr-democracy-gene-editing-inclusive-deliberation/ [17-7-2021].

[11] *Ibid.*

[12] Cf. Pellegrino in P. Martinho da Silva, «Bioethics: And Politics», in H. ten Have (ed.), *Encyclopedia of Global Bioethics*, Springer International Publishing, Cham 2016, 291–296, 296.

on procedures more than on ethics. However, ethics is important because it asks the difficult and fundamental questions. The European Group on Ethics in Science and New Technologies stated that ethics for the future needs to be "an ambitious, proactive, daring and public ethics, an ethics that asks the hard, painful or tedious questions.".[13] This is the kind of ethics that "addresses the foundations of the practices and institutions that are responsible for the phenomena that it considers to be problematic."[14]

Many other specialists can and hopefully will improve global governance procedures and mechanisms. Bioethicists should search for, show the goodness of, and spread ethical truths. This is not a simple task. First, because moral truth is not evident in novel issues such as HGE, the search for ethical truth might take a long time and be filled with nuances. Second, it is not enough that bioethicists find clarity regarding the moral applications of HGE. Bioethical reasoning must reach the other key agents of global governance and shape their predominant values.

8.2 Bioethics and the Key Agents of the Governance of HGE

Governance includes multiple stakeholders. In this work, four groups are referred to as the key agents in the global governance of HGE: researchers, policymakers, funders, and the public. This choice was based on the main governance elements resulting from the RTA of the selected H + and WHO texts: research, policies, funds, and public debate. Clearly there are other agents, but it is these agents which were addressed in a relevant way by both H + and WHO texts.

8.2.1 Researchers

Let's imagine the following scenario in which researchers create a new biotechnology. The researchers then send this new technology to an ethics committee for approval or disproval. Following the spread of this new technology, laws related to it are enacted. Policy-making follows ethical reflection. Laws are created nationally, then internationally. People abide by these laws.

This scenario fails not only because it is idealistic, but also because it is linear and places ethics in pre-established moments. In reality, researchers have ideas, desires and intentions when they research and when they give importance to a finding. Their ethical ideas influence their research. Ethical committees are formed by diverse members, who have diverse ethical educational backgrounds and diverse

---

[13] European Group on Ethics in Science and New Technologies, *Values for the Future: The Role of Ethics in European and Global Governance*, Publications Office of the European Union, Brussels 2021, in https://ec.europa.eu/info/files/values-future-role-ethics-european-and-global-governance_en [16-11-2021],16.

[14] *Ibid.*

values. Policymakers follow a set of values as well. People across the globe who break or respect the law are also influenced by their personal values.

Researchers, as any other moral agent, might find a way to move around the law if they do not support the reasons behind it. For example, in the case of the first three-parent baby, where an embryo was produced and the child was born with the DNA of three people through mitochondrial replacement therapy (MRT), sometimes called mitochondrial donation. This procedure was done in Mexico, even though neither the parents nor the medical staff were Mexican. The main researcher said the experiment happened there because in Mexico there is no law against this specific procedure.[15] Researchers' choices might get complicated even if an unethical or controversial experiment is not explicitly prohibited. Cases such as He Jiankui's CRISPR babies challenge the ability of science to regulate itself and the world's trust in it.[16]

Researchers cannot escape their duty to search for and update their ethical training. They cannot wash their hands, leaving ethical reasoning to ethics committees, policymakers, or the final users of their inventions. Bioethicists can help researchers to integrate some ethical concepts and reasonings, such as why the intention does not justify the means, how ethics and law are not the same thing and an action can be legal and not ethical, the issue of technical possibilities versus moral goodness, dual-use dilemmas, protection of the vulnerable, and global health justice.

8.2.2 Policymakers

Legality is not the same as morality. A situation can be legal in a given place and time, and still be profoundly unethical. If policies are made based only on the consensus of the majority, this might lead to unethical cases, such as slavery in the past. Policymakers have a great responsibility to integrate moral values into their decisions.

The idea of malleability and morphological freedom, along with other liberties and rights claimed by transhumanists, poses many challenges to policymakers. For instance, the U.S. Transhumanist Party published the *Transhumanist Bill of Rights* in 2015. It was delivered to the U.S. Capitol on December 14th, 2015. This first version of the *Transhumanist Bill of Rights* contained the following six articles:

[15] Cf. J. Paullier, «¿Por qué la concepción del bebé de 'tres padres' se realizó en México?», *BBC News* (2016), in https://www.bbc.com/mundo/noticias-37491942 [25-1-2020].

[16] Cf. H. T. Greely, «How Should Science Respond to CRISPR'd Babies?», *Issues in Science and Technology* XXXV/3 (2019), 32–37, in https://issues.org/how-should-science-respond-to-crisprd-babies/ [18-12-2019], 37.

> **Article 1.** Human beings, sentient artificial intelligences, cyborgs and other advanced **sapient life forms** are entitled to **universal right**s of **ending involuntary suffering**, making personhood **improvements**, and achieving an **indefinite lifespan** via science and technology.
>
> **Article 2.** Under penalty of law, no cultural, ethnic, or religious perspectives influencing government policy can impede **life extension science**, the health of the public, or the possible maximum amount of life hours citizens possess.
>
> **Article 3.** Human beings, sentient artificial intelligences, cyborgs and other advanced sapient life forms agree to uphold **morphological freedom** – the right to do with one's physical attributes or intelligence (dead, alive, conscious, or unconscious) whatever one wants so long as it doesn't hurt anyone else.
>
> **Article 4.** Human beings, sentient artificial intelligences, cyborgs and other advanced sapient life forms will take every reasonable **precaution to prevent existential risk**, including those of rogue artificial intelligence, asteroids, plagues, weapons of mass destruction, bioterrorism, war, and global warming, among others.
>
> **Article 5.** All nations and their governments will take all reasonable measures to embrace and fund **space travel**, not only for the spirit of adventure and to gain knowledge by exploring the universe, but as an ultimate safeguard to its citizens and transhumanity should planet Earth become uninhabitable or be destroyed.
>
> **Article 6. Involuntary ageing shall be classified as a disease.** All nations and their governments will actively seek to dramatically extend the lives and improve the health of its citizens by offering them scientific and medical technologies to overcome involuntary ageing.[17]

The current version, (3.0), was adopted by the U.S. Transhumanist Party in 2018. This third *Transhumanist Bill of Rights* passed from 6 to 43 articles in only three years.[18] This third version has many parallels to the *Universal Declaration of Human Rights.*[19] But the differences are enormous, starting with the subjects of rights and duties. In the Preamble, *The Transhumanist Bill of Rights 3.0* states that the term "sentient entities" encompasses:

- Human beings, including genetically modified humans.
- Cyborgs.
- Digital intelligences.
- Intellectually enhanced, previously non-sapient animals.
- Any species of plant or animal which has been enhanced to possess the capacity for intelligent thought.
- Other advanced sapient life forms.[20]

---

[17] U.S. TRANSHUMANIST PARTY, «Transhumanist Bill of Rights», 2015, *H + Pedia*, in https://hplus pedia.org/wiki/Transhumanist_Bill_of_Rights [9-1-2023]. The bold is mine.

[18] Cf. U.S. TRANSHUMANIST PARTY, «Transhumanist Bill of Rights – Version 3.0», 2018, in https://transhumanist-party.org/tbr-3/ [14-4-2021].

[19] Cf. UN, «Universal Declaration of Human Rights», 1948, in https://www.ohchr.org/en/human-rights/universal-declaration/translations/english [13-7-2019].

[20] Cf. U.S. TRANSHUMANIST PARTY, «Transhumanist Bill of Rights – Version 3.0», 2018, in https://transhumanist-party.org/tbr-3/ [14-4-2021].

These ideas present a challenge to the current interpretation of legal subjects. For instance, some felt it was necessary to clarify Article 2 of the *European Convention on Human Rights*, which protects *everyone's* right to *life*, referring to "everyone" as humans, and "life" as human life.[21]

In addition to attempting a broadening of the legal persons, the *Transhumanist Bill of Rights 3.0* also presents new rights, following the liberties claimed by transhumanists. For example, morphological freedom: "Article X. Sentient entities agree to uphold morphological freedom—the right to do with one's physical attributes or intelligence whatever one wants so long as it does not harm others."[22] Along with reproductive freedom: "Article XII. All sentient entities are entitled to reproductive freedom, including through novel means such as the creation of mind clones, monoparent children, or benevolent artificial general intelligence."[23]

In 2018, a group of jurists gathered, noting that Transhumanism already introduces a significant challenge for politics and law today.[24] "For once nature is removed from human affairs, everything becomes conventional, that is, accidental: morality, law, family, politics, life itself."[25] From lawyers' perspectives, transhumanist ideas and rights bring forth many implications for civil law, family law, public law, property law, and criminal law.[26]

Bioethicists can contribute to policymakers producing relevant literature that can inform and support ethical regulations. This literature may include ethical reasoning, promoting the understanding that consensus does not constitute morality, and showing how unethical regulations have consequences for individuals and society.

8.2.3 Funders

As shown in both Humanity + and WHO texts, funds influence and make possible the global governance of new technologies. Knowing that funding is an important element regarding the research, decisions, and implementation of new technologies, the question is: who decides the order of priorities for the allocation of funds? Following what criteria?

---

[21] Cf. D. KORFF, «The Right to Life: A Guide to the Implementation of Article 2 of the European Convention on Human Rights», (2008), in https://papers.ssrn.com/abstract=1288555 [8-11-2022], 8.

[22] U.S. TRANSHUMANIST PARTY, «Transhumanist Bill of Rights – Version 3.0», 2018, in https://transhumanist-party.org/tbr-3/ [14-4-2021].

[23] *Ibid.*

[24] Cf. M. AYUSO (ed.), *¿Transhumanismo o Posthumanidad? La política y el derecho después del humanismo*, Prudentia Iuris, Marcial Pons Ediciones Jurídicas y Sociales, Madrid 2020.

[25] J. F. SEGOVIA, «La progresiva destrucción de la naturaleza y la naturaleza humana», in M. AYUSO (ed.), *¿Transhumanismo o Posthumanidad? La política y el derecho después del humanismo*, Prudentia Iuris, Marcial Pons Ediciones Jurídicas y Sociales, Madrid 2020, 49–82, 74. The translation is mine. The original follows: 'Porque una vez que se remueve la naturaleza en los asuntos humanos, todo se vuelve convencional, es decir, accidental: moral, derecho, familia, política, la vida misma.'.

[26] Cf. *Ibid.*, 75–77.

In a world filled with all sorts of problems, Zoltan Istvan, founder of the U.S. Transhumanist Party, affirms that the priority is to create technologies to stop ageing and disease. In his own words: "This is the main goal of transhumanists, to overcome death. We can worry about many other things later, like social equality, ending wars, and ending poverty. But first we must strive to create technologies that stop the world's greatest killers: aging and disease."[27] Those who agree with Zoltan are not the same ones who think they can die in a conflict, war or famine before ageing. Istvan's comment lacks recognition of the link between poverty-hunger-disease-death, and the link between social inequality-violence-death.

Gene editing for health is promising,[28] and certainly merits to be funded. However, funds are limited and should be wisely distributed in an economy that wants humanity to flourish and thrive. Currently, what are the true priorities for the allocation of funds for research? Henk ten Have, in *Global Bioethics: an Introduction*, discusses the 10/90 gap. The Global Forum for Health Research estimated that less than 10% of the $55 billion global spending on health research is devoted to diseases that account for 90% of the global burden of disease. Together, pneumonia, diarrhoea, tuberculosis, and malaria accounted for 20% of the disease burden in the world, yet received less than 1% of public and private funds for health research.[29] "Disparity is the result of economic as well as policy decisions. It is not profitable to develop new drugs for poor populations that cannot afford them. There also is no urgency since these populations usually cannot put any pressure on policy-makers."[30]

Will this gap be aggravated by the increasing funding destined for gene editing research? And regarding the funds allocated to gene editing, will they prioritise conditions that affect many people around the globe, or will they be centred on rare diseases?

Currently, some "superstar genes" (such as *TP53*, *TNF* and *EGFR*) have become the subject of hundreds of publications a year, while most other genes receive little attention. Intense study is surely justified for genes that have profound biological importance. For example, *TP53* is crucial for cell growth and death, leading to cancer when inactivated or altered.[31] However, the focus on this gene is also

---

[27] Z. ISTVAN, «Expectations and Apprehensions on Transhumanism», *FIfF-Kommunikation 2/2016* 2/16 (2016), 19–23, 21.

[28] The article presents a review of recent advances of the three major genome editing technologies (ZFNs, TALENs, and CRISPR/Cas9) for therapy. Cf. H. LI ET AL., «Applications of genome editing technology in the targeted therapy of human diseases: mechanisms, advances and prospects», *Signal Transduction and Targeted Therapy* 5/1 (2020), 1–23, in https://www.nature.com/articles/s41392-019-0089-y [9-5-2022].

[29] Cf. H. TEN HAVE, *Global Bioethics: An Introduction*, Routledge, New York 2016, 62.

[30] *Ibid.*

[31] Cf. A. J. GATES ET AL., «A wealth of discovery built on the Human Genome Project—by the numbers», *Nature* 590/7845 (2021), 212–215, in https://www.nature.com/articles/d41586-021-00314-6 [21-5-2021], 212.

heavily influenced by what is called preferential attachment[32] in network science. It is easier to secure funding, mentorship, tools, and citations for further research on already highly published genes. It is a safe bet. Consequently, although many genes are highly studied, only a few out of many are highly studied. For instance, by 2017, 22% of gene-related publications referenced only 1% of genes.[33] The article *A wealth of discovery built on the Human Genome Project—by the numbers* affirms that the number of new yearly publications focusing on a given gene is linearly proportional to the size of previous literature on it.[34] Authors argue that the next challenge is to disentangle the motivations for what gene gets studied next. "Are researchers putting money, time and effort into what is most important or urgent, or into more of the same because that will reliably win grants and plaudits?".[35]

How can Bioethics help to wisely set priorities for the allocation of funds? A possible way is to separate priorities for public funding from priorities for private funding. Public funding is derived from the citizens' taxes and should be invested in important issues that can improve the life of general society. The percentage of public investment should be proportional to public needs. For instance, if malaria is a considerable issue in a given country, it would be unfair to allocate the majority of health research funding to a rare disease.

However, in practice, societies have not waited until hunger was solved in all parts of the globe to invest in planes or satellites, for example. While public funding should be proportional to a given community's needs, individuals have the freedom to invest in different areas according to their interests. Yet, individuals should be socially aware, and ensure that the objective of their investment will not harm the community, and will hopefully improve some aspect of the life of others. Private funding can support a community by increasing the range of funded causes, which public funding would not be able to fairly cover.

Bioethicists can contribute to funders by providing criteria to determine the priorities for the allocation of funds. It is in the funders' hands to incentivise and direct creative people to solve current problems in the best way possible.

8.2.4 The Public

Both Humanity + and WHO texts give great importance to public debate. Why is public engagement so desired? Is public debate realistic? What is the role of

---

[32] Cf. A. L. BARABASI - R. ALBERT, «Emergence of scaling in random networks», *Science* 286/5439 (1999), 509–512.

[33] Cf. A. J. GATES ET AL., «A wealth of discovery built on the Human Genome Project—by the numbers», *Nature* 590/7845 (2021), 212–215, in https://www.nature.com/articles/d41586-021-00314-6 [21-5-2021], 212.

[34] Cf. *Ibid.*

[35] *Ibid.*, 214.

public debate? This section addresses the feasibility and the goals, some specific initiatives, and some challenges of public engagement regarding the HGE debate.

Is societal consensus a feasible goal? Alta Charo, co-chair of the NASEM committee that issued the 2017 report on HGE,[36] and member of the organising committee for the Second and Third International Summits on HGE, wrote that a broad social consensus is impossible. "Some critics [of the 2018 summit statement on human genome editing] weaponized the language of 'broad social consensus' used by the first summit's organizers, calling for an indefinite moratorium until such consensus could be reached, without describing what that consensus might look like. Certainly, a global consensus (by majority? calculated by polling? calculated by voting?) is simply impossible."[37]

In contrast, F. Baylis, a member of the WHO Expert Advisory Committee, and member of the organising committee for the First and Third International Summits on HGE, expressed her disagreement: "The suggestion here [by Alta Charo] is that broad societal consensus is an unattainable ideal. And yet, the many references to He having violated international consensus belie this claim. Moreover, this perspective ignores recent efforts to address the meaning and scope of broad societal consensus."[38]

The feasibility of a social consensus depends on its meaning and its goal. Sometimes public participation seems to be aimed at social consensus, and some other times it seems that the expectations are rather to engage in a dialogue to enrich perspectives. What is the meaning and the scope of societal consensus? The general idea is that public participation is not aiming for unanimity. Baylis says: "As a staunch proponent of broad societal consensus, I have tried to explain that this is not about unanimity, but that it also doesn't collapse into majority rule."[39] According to Baylis, "what matters (and perhaps matters most) with broad societal consensus is the journey or the process." Specifically regarding the public debate on heritable human gene editing, Baylis writes: "My bottom line is that the human genome, metaphorically speaking, belongs to all of us. We should all have a say in whether to proceed with heritable genome editing."[40]

Similarly, Pete Mills, one of the authors of the 2018 Nuffield report,[41] writes about the dynamic relationship among three venues in which discourses on science and technology are played out: the Republic of Science, the Halls of Justice, and the Public Sphere. "We can recognize certain sorts of pathologies that come from the dynamics

---

[36] Cf. NATIONAL ACADEMIES OF SCIENCES, ENGINEERING, AND MEDICINE, «Human Genome Editing: Science, Ethics, and Governance», The National Academies Press, Washington, D.C. 14 February 2017, in https://doi.org/10.17226/24623 [5-7-2021].

[37] R. Alta Charo in F. BAYLIS, «Human Genome Editing: Our Future Belongs to All of Us», *Issues in Science and Technology* XXXV/3 (2019), 42–44, in https://issues.org/wp-content/uploads/2019/04/Baylis-Human-Genome-Editing-SP19-ISSUES-42-44.pdf [18-12-2019], 44.

[38] *Ibid.*

[39] *Ibid.*

[40] *Ibid.*

[41] Cf. NUFFIELD COUNCIL ON BIOETHICS, *Genome editing and human reproduction: Social and ethical issues*, Nuffield Council on Bioethics, London 2018, in https://www.nuffieldbioethics.org/publications/genome-editing-and-human-reproduction [20-4-2018]

and imbalances of power among these three venues. For example, ignoring the third venue can lead to elitism; and ignoring the first can lead to populism."[42]

Furthermore, in the *Geneva Statement*, a group of relevant authors state that the decision of whether we should and how to proceed with heritable human gene editing is not under the authority of a small but influential group: "(…) a small but vocal group of scientists and bioethicists now endorse moving forward with heritable human genome editing. They have taken it as their task to decide how we might proceed toward altering the genes of future children and generations. In fact, the question at hand is whether to proceed at all."[43]

The flip side raises another question: should scientists, ethicists, policymakers, and health staff do what the general people want? For example, the *2018 Nuffield Report* says that prospective care providers want to *have a child, genetically related to them*, they want the child they have *not to have a specified condition*, and/or they want the child *to have a specified characteristic*.[44] It seems that the idea behind this notion is that scientists should give people what they want. This becomes even more complicated considering that the same report says that public norms are subject to change: "Public moral norms can evolve (e.g. attitudes to sexual orientation) and respond to experience (e.g. norms relating to IVF and embryology). (…) Changes can come from any direction: changes in statute law over time, for example, both reflect and affect such norms."[45]

If the goal of public debate surrounding the global governance of HGE is to determine the moral truth, this is not only difficult but also dangerous. Public opinion is not a guide to what is right or wrong. Think of misogynistic or racist societies, where there is social consensus for these very unfair practices. Societal consensus does not guarantee ethics and justice, and it surely does not avoid the consequences of unethical but socially accepted choices. Public debate is not to make decisions based only on what people want. Although public input is important regarding decisions that affect the health of a population, the consensus of the majority is not enough to determine its ethics, nor to protect the rights of minorities and the socially vulnerable.

The first possible aim of public debate is societal consensus, which appears to be difficult to measure, and even dangerous if taken as the determinant of morality. The second possible goal of public debate is to mutually enrich perspectives. In this second scenario, considering the input of diverse citizens can positively add to governance. Public engagement is useful when looking at other perspectives,

---

[42] Cf. P. MILLS, «Three Venues for Discussing Human Gene Editing», *Issues in Science and Technology* XXXV/3 (2019), 38–41, in https://issues.org/three-venues-for-discussing-human-gene-editing/ [18-12-2019], 41.

[43] R. ANDORNO ET AL., «Geneva Statement on Heritable Human Genome Editing: The Need for Course Correction», *Trends in Biotechnology* 38/4 (2020), 351–354, in https://www.sciencedirect.com/science/article/pii/S0167779919303178 [4-3-2021], 352.

[44] Cf. NUFFIELD COUNCIL ON BIOETHICS, *Genome editing and human reproduction : Social and ethical issues*, Nuffield Council on Bioethics, London 2018, in https://www.nuffieldbioethics.org/publications/genome-editing-and-human-reproduction [20-4-2018], 23.

[45] *Ibid.*, 53.

as individual and peer perspectives are limited and can miss important points. For instance, we are all biased by our cultures. For example, knowing the sex of an unborn child might be good within some cultures, but impose negative pressure in cultures that favour a male.

To achieve the goal of public engagement, the organisers of the dialogue must remember that it is a *dialogue*, not a communication or an information section. The goal is to mutually enrich each other, not to teach and convince the public that this technology should or should not develop in a pre-established way. We tend to be more focused on communicating than listening. People calling for public dialogue might be more interested in creating public trust (as it affects policy making, market, and funding), or even checking public dialogue off the to-do list, than seeking unbiased input from the public. It is important to be attentive to avoid biased communication and engage in a fruitful dialogue, one that gives and receives inputs.

If the aim is to enrich perspectives, public debate is not a matter of the number of people participating, but of the diversity among participants. The dialogue must include different stakeholders. The type of participants should vary according to the objective of the specific dialogue. Below is a list of examples of people that should be included in the HGE public debate:

- Directly affected people (such as patients and their relatives).
- Healthcare professionals.
- People with disabilities.[46]
- People of different ages.
- People of different sexes.
- People from different cultural backgrounds.
- Religious representatives.[47]

---

[46] We usually fail to predict the impact of disabilities. Some authors looked at possible sources of error in both the happiness reports of patients with chronic illness or disability and the happiness predictions of healthy people asked to imagine the same illnesses and disabilities were collected and analysed. The available evidence suggested that "whereas patients misreport their well-being, healthy people also mispredict the emotional impact that chronic illness and disability will have on their lives." P. A. UBEL ET AL., «Misimagining the unimaginable: the disability paradox and health care decision making», *Health Psychology: Official Journal of the Division of Health Psychology, American Psychological Association* 24/4S (2005), S57-62, S57.

[47] More than 80% of the global population belong to some religious tradition. Cf. PEW RESEARCH CENTER'S RELIGION & PUBLIC LIFE PROJECT, «The Global Religious Landscape», 2012, *Pew Research Center*, in https://www.pewresearch.org/religion/2012/12/18/global-religious-landscape-exec/ [7-9-2022].

Yet, religious perspectives are often not considered in the deliberation of bioethical issues. Some have explored the phenomenon of religious perspectives leaving the bioethical debate. Cf. J. H. EVANS, *Playing God? Human Genetic Engineering and the Rationalization of Public Bioethical Debate*, University of Chicago Press, Chicago 2002; Cf. J. THAM, *Secularization of Bioethics-A Critical History*, Ateneo Pontificio Regina Apostolorum, Rome 2007.

The following is an example of a project considering religious perspectives on bioethical issues. Cf. J. THAM, «生命倫理學:跨文化研究: Bioethics: Cross-Cultural Explorations», *International Journal of Chinese & Comparative Philosophy of Medicine* 20/2 (2022), 13–37, in

- Advocacy groups related to the topic.
- Traditionally underrepresented groups (such as indigenous peoples, minority ethnic groups or faiths).

Regarding initiatives for public engagement, it should be noted that public debate on bioethical topics is not something new. For instance, Baylis offers a summary of the cornerstones that have occurred in public debate. The author starts by noting Mary Shelly's *Frankenstein* (1818) and creates a list until reaching the Call for a Moratorium and the creation of the WHO Advisory Committee (2019).[48] The following present some initiatives that can be taken regarding public engagement in the current gene editing debate:

- Research on the public views of HGE, for example:
  - Public Acceptability of Gene Therapy and Gene Editing for Human Use: A Systematic Review.[49]
  - Public views on polygenic screening of embryos.[50]
- Organisations fostering public participation such as:
  - Center for Genetics and Society[51]: a non-profit organisation that aims to foster public dialogue, by gathering relevant reports, news and articles on the topic.
  - Global Observatory for Genome Editing[52]: built up after a call from Jasanoff and Hurlbut.[53] The Global Observatory aims to foster international, interdisciplinary, and cross-sectoral dialogue about gene editing. It maps both dominant and marginalised ideas of care present in discourses on HGE, convenes conversations on more expansive ideas of human flourishing, and produces papers on policy development.

https://ejournals.lib.hkbu.edu.hk/index.php/ijccpm/article/view/2238 [9–1-2023]; Cf. UNESCO CHAIR IN BIOETHICS AND HUMAN RIGHTS, «Bioethics, Multiculturalism and Religion», in https://www.unescobiochair.org/bioethics-multiculturalism-and-religion-3/ [15–2-2023]. The project publication that best addresses HGE is J. THAM - C. DURANTE - A. G. GÓMEZ (edd.), *Interreligious Perspectives on Mind, Genes and the Self: Emerging Technologies and Human Identity*, Routledge Science and Religious Series, Routledge, Oxon and New York 2019.

[48] Cf. F. BAYLIS, *Altered Inheritance*, 84–85.

[49] Cf. J. DELHOVE ET AL., «Public Acceptability of Gene Therapy and Gene Editing for Human Use: A Systematic Review», *Human Gene Therapy* 31/1–2 (2020), 20–46, in https://www.liebertpub.com/doi/10.1089/hum.2019.197 [5-7-2021].

[50] Cf. M. N. MEYER ET AL., «Public views on polygenic screening of embryos», *Science* 379/6632 (2023), 541–543, in https://www.science.org/doi/10.1126/science.ade1083 [20-2-2023].

[51] Cf. CENTER FOR GENETICS AND SOCIETY, «Welcome to CGS», in https://www.geneticsandsociety.org/ [2-7-2021].

[52] Cf. GLOBAL OBSERVATORY FOR GENOME EDITING, «Global Observatory», 2020, in https://global-observatory.org/about/ [15-7-2021].

[53] Cf. S. JASANOFF - J. B. HURLBUT, «A global observatory for gene editing», *Nature* 555/7697 (2018), 435–437, in https://www.nature.com/articles/d41586-018-03270-w [24-5-2022].

  - ARRIGE[54]: a non-profit organisation that aims to promote responsible uses of gene editing, with a focus on the global South, including South Asia, Africa, and South America.
- Forums, assemblies, debates, and workshops such as:
  - Workshop on public dialogue on genomic medicine, promoted by the Council of Europe.[55]
  - *CRISPRcon*[56]: creates a unique forum in which a broad selection of diverse voices come together to discuss future gene editing technologies across a variety of applications in agriculture, health, conservation and more.
  - *Global Citizens' Assembly on Genome Editing*[57]: aims to contribute to meaningful public engagement at the global level, enacted the world's first in-person Global Citizens' Assembly.

The WHO framework for the governance of HGE offers a box with good practices in public education, engagement and empowerment. However, it is generic and only mentions topics such as public education and social media/advertising campaigns, discussion-based tools, shared decision-making tools, and the values of openness, transparency, honesty and accountability.[58] It does not recommend any methodology for public engagement in the HGE debate.

Each initiative has its advantages and its flaws. "Public forums, surveys, focus groups, and other methods for gathering public input may complement polling, but also suffer from insufficient reach, self-selection of participants, and inattention to differential power dynamics in presumed consensus building or democratic processes."[59] Since no methodology is perfect, perhaps the better recommendation is to use different methodologies when trying to respond to the wide range of challenges regarding public engagement.

The challenges of public debate involve public education, the selection of representatives for the dialogue, funding, the selected methodologies for the dialogue,

---

[54] Cf. «ARRIGE – Association for Responsible Research and Innovation in Genome Editing», in https://www.arrige.org/ [7-3-2023].

[55] Cf. Council of Europe, «Workshop on public dialogue on genomic medicine», 2022, *Human Rights and Biomedicine*, in https://www.coe.int/en/web/bioethics/workshop-public-dialogue-genomic-medicine [10-11-2022]; Cf. M. Bale, «Report: Workshop on Public Dialogue on Genomic Medicine», Council of Europe, Strasburg 25 April 2023, 20, in https://rm.coe.int/cdbio-2023-1-rapporteur-report-on-public-dialogue-genomic-medicine-e/1680aae608 [1-5-2023].

[56] Cf. «CRISPRcon - Conversations on Science, Society and the Future of Gene Editing», *CRISPRcon*, in https://crisprcon.org/ [5-3-2023].

[57] Cf. «Global Citizens' Assembly on Genome Editing», *Globalca*, in https://www.globalca.org [7-3-2023].

[58] Cf. WHO Expert Advisory Committee on Developing Global Standards for Governance and Oversight of Human Genome Editing, *Human genome editing: a framework for governance*, WHO, Geneva 2021, in https://www.who.int/publications/i/item/9789240030060 [13-7-2021], 15.

[59] J. Halpern et al., «Societal and Ethical Impacts of Germline Genome Editing: How Can We Secure Human Rights?», *The CRISPR Journal* 2/5 (2019), 293–298, in https://www.liebertpub.com/doi/full/10.1089/crispr.2019.0042 [4-3-2021].

navigating the lack of consensus, the use of information and communication, accountability, and the practical use of public inputs.

Bioethicists can contribute to public dialogue by clarifying the goal of public engagement, creating awareness that societal consensus is not to determine morality, and emphasising the fruit of a mutually enriching exchange of perspectives. Furthermore, Bioethicists can ensure that the focus is on dialogue, not only on education or communication from the organisers (attention to biased communication). It also recommends the diversification of the participants, and the use of diverse methodologies since no methodology is perfect.

8.3 Values Supporting Ethical Barriers in the Public Debate

After addressing the main agents of the global governance of HGE, values should be addressed in more detail, as they guide ethical decisions. The common values found in the RTA (See Chap. 3)[60] were divided into three groups: *respect for individual rights-autonomy in health and reproduction*, *protection of people who cannot express their wishes*, and *equality-solidarity*. In order to understand the historical and sociological role of predominant values in the discussions on HGE, a broader context of values will be given.

In the book *The Human Gene Editing Debate*, John Evans[61] affirms that public bioethical debates are generally set up as slopes with the most acceptable at the top and the most abhorrent at the bottom. He gives the example of the social acceptability of abortion, with the terrain of the slope typically being defined by foetal age and the reasons the woman wants the abortion. At the top of the slope is abortion to save the pregnant woman from imminent death, and at the bottom is infanticide for no specific reason. Most people want to be at the top and do not want to fall to the bottom. Evans also gives the example of euthanasia. At the top of the slope is euthanasia for fully competent adults who are going to imminently die a painful death due to disease, and

[60] The full Reflexive Thematic Analysis (RTA) can be consulted in the published doctoral dissertation excerpt L. SANTOS, *Common Themes Between Humanity + Original Documents on Transhumanism and WHO Documents on the Global Governance of Human Genome Editing*, Ateneo Pontificio Regina Apostolorum, Rome 2023.

[61] John H. Evans is the Tata Chancellor's Chair in Social Sciences, Professor of Sociology, Associate Dean of Social Sciences and Co-Director of the Institute for Practical Ethics at the University of California, San Diego. He was one of the 22 members of NASEM committee that worked on the HGE 2017 report. Cf. NATIONAL ACADEMIES OF SCIENCES, ENGINEERING, AND MEDICINE, «Human Genome Editing: Science, Ethics, and Governance». He said there was consensus on the publication, but not in every aspect of the report. In his book, Evans writes his own opinion. He says the barriers that can be built from the NASEM report were not rejected by the group, but he had rather failed to consider the possibility of these barriers until long after the NASEM report was completed. Cf. J. H. EVANS, *The Human Gene Editing Debate*, 167 (note 23).

at the bottom is involuntary death for authorities that conclude one should no longer live.[62]

Evans explains that there are many reasons why public bioethical debates tend to be structured as slopes. Firstly, the origin of the public bioethical debate during the 1960s and 1970s was in the overreach of scientists and physicians, and the public calling for limits to avoid a dystopian future. The public bioethical debate was invented to place limits, which encourages arguments that function on a slope: "Above here is acceptable, below here is unacceptable." Secondly, since bioethical debates are ultimately input for regulation, they conform to the nature of law in liberal democratic societies: citizens are free to do anything they want as long as there is no law against it. Laws should *only* be created for good reasons, such as to prevent an action that might harm someone.[63]

Evan's book presents the mechanism of barriers to avoid ethical slippery slopes. "Slipping is when the consensus of what is ethical in the debate moves to the next step down the slope."[64] Evans affirms that "a barrier is strong if all the values that support it have a positive evaluation of the actions upslope and a negative evaluation of the actions downslope."[65] The slipping factor is due to the absence of a barrier between the meritorious step and the one below it. In addition to a possible change of values, there are also two reasons for the weakening and the dissolution of a barrier: continuity vagueness and similarity vagueness. Continuity vagueness is when there can be no argument in the difference between, say, 17 and 18. Similarity vagueness is "where distinction is not possible because of the only distinction that can be made between the steps does not hold ethical weight (e.g., a disease of the liver vs. a disease of the kidneys)."[66]

In synthesis, Evans' slippery slope in bioethical debate theory implies that:

- Bioethical debate is structured as a slope (ideal top versus dystopian bottom).
- Barriers supported by values are set to avoid slipping towards the bottom of the slope.
- Barriers fall due to weak construction (continuity, similarity vagueness), or change of values.
- Slipping occurs when the consensus of what is ethical in the debate moves to the next step down the slope.

Concerning the HGE debate, Evans has identified some past, present, and possible future barriers. These barriers are summarised in the following Table 8.1.

[62] Cf. J. H. Evans, *The Human Gene Editing Debate*, Oxford University Press, New York 2020, 11–13.
[63] Cf. *Ibid.*
[64] *Ibid.*, 145.
[65] *Ibid.*, 38.
[66] *Ibid.*

Table 8.1 Barriers in the human gene editing debate

Era	Barriers
Before CRISPR	Disease-enhancement
	Somatic-germline
CRISPR era	Prevalent variant
	Median trait
Future (possible barriers)	Goals of medicine
	Family genes
	Boundary of humanity
	Liberal eugenics
	Scientific reality

Adapted from J. H. Evans' book The Human Gene Editing Debate[67]

Each of these barriers was built upon values and designed with certain strengths. Unfortunately, each barrier also has weaknesses. The next series of tables synthesises Evan's barriers in the HGE debate, the values supporting each barrier, and their weakness.[68]

8.3.1 Before CRISPR

See Tables 8.2 and 8.3

These first two barriers (somatic/germline and disease/enhancement) were built in the 70s. Back then, disease was not considered as something problematic to define, and did not coincide with all the traits that eugenics thought were bearing down on humanity. Enhancement came to mean all those traits of the eugenicists' dreams: intelligence, artistic ability, and good character.[69] These barriers were strong in the 70s and 80s. However, due to their above-mentioned weaknesses, by the end of the 20th Century, there was a lack of strong barriers. The value of non-maleficence still resisted supporting both barriers. Scientists would not go beyond, because to do so was not yet demonstrated as safe.

These barriers could not persevere during the CRISPR era. As safety increased with CRISPR, the somatic-germline barrier could not hold. In 2015, a Chinese scientist edited a human embryo (not implanted),[70] and in 2018 another Chinese researcher claimed the first gene-edited babies were born.[71] Other barriers were attempted to be in place.

---

[67] Cf. *Ibid.*, 38–148.

[68] Cf. *Ibid.*

[69] Cf. *Ibid.*, 31.

[70] Cf. P. Liang et al., «CRISPR/Cas9-mediated gene editing in human tripronuclear zygotes», *Protein & Cell* 6/5 (2015), 363–372, in https://doi.org/10.1007/s13238-015-0153-5 [11-11-2022].

[71] Cf. M. Marchione, «Chinese researcher claims first gene-edited babies», *AP NEWS* (2018), in https://apnews.com/article/4997bb7aa36c45449b488e19ac83e86d [19-12-2020].

Table 8.2 Disease-enhancement barrier

Disease-enhancement barrier
Strength *Medical profession*: beneficence is interpreted as healing suffering from disease, which is interpreted as bodily dysfunction and not as treating personal desires
Supporting values *Beneficence:* Above the barrier: beneficence is interpreted as relief from suffering Below the barrier: non-beneficence is interpreted as egotism and wanting to be better than everyone else *Following God's will:* Above the barrier: repairing diseases that were a result of the Fall Below the barrier: trying to improve on God's design for the species
Weakness *Problems with the concept of disease:* serious disease vs minor disease vs genetic variation; the possibility to foresee predisposition to diseases; "disease as monogenic and enhancement as polygenic" was not true; the autonomy of parents practising eugenics on a personal level; "disease as deviation from *normalcy*?"

Adapted from J. H. Evans' book *The Human Gene Editing Debate*[72]

Table 8.3 Somatic-germline barrier

Somatic-germline barrier
Strength *Specific target*: individual or species (which could not be met but was an abstraction) *Clear concepts:* somatic or germline. But lack of adequate ethical force
Supporting values *Non-maleficence*: Above the barrier: somatic interventions. Below the barrier: the worst-case scenario was the death of the person *Humility*: Above the barrier: caution, limits. Below the barrier: hubris, humans redesigning the species *Social equality:* Above the barrier: people as individuals. Below the barrier: deciding which traits held by which groups should be the future of species *Respecting nature or God's will*: Above the barrier: respect was interpreted as non-intervention, only fixing problems. Below the barrier: human control identified with eugenics, highly repulsive at the moment
Weakness *A change in participants and their values* (theologians-bioethicists, 4 principles) *Emphasis on beneficence and autonomy* undermines the germline barrier *Technology changes the terrain* (IVF, PGD), and new barrier downslope (germline selection, then germline modification…) *The focus switches to technical questions* of whether somatic gene therapy could be done safely

Adapted from J. H. Evans' book *The Human Gene Editing Debate*[73]

[72] Cf. *Ibid.*, 40–66.

[73] Cf. *Ibid.*, 40–66.

8.3.2 CRISPR Era

See Tables 8.4 and 8.5

Table 8.4 Prevalent variant barrier

Prevalent variant barrier
Strength *Prevalent trait*: widespread in a particular area at a particular time
Supporting values *Justice:* interpreted as access to what is the most common in a population
Weakness *The concept of "prevalent"* is subjected to continuity vagueness. Twisting to the most prevalent avoids continuity vagueness as only one variant is "the most" *The concept of "area"*: region, country, world? Risk of population shopping *This will hold as long as technology is limited to monogenic modification* *Beneficence and autonomy* threaten the supremacy of justice. Autonomy*:* this barrier is below what many people want because it accepts HGE use beyond "serious" disease, germline modifications, and upgrading or downgrading to a prevalent variant

Adapted from J. H. Evans' book *The Human Gene Editing Debate*[74]

Table 8.5 Median trait barrier

Median trait barrier
Strength *The middle person in a population is ordered along some scale*
Supporting values *Justice*: interpreted as equality of opportunities
Weakness *It means slowly upgrading humanity* *A different interpretation of the value of justice*: societally useful inequalities, making some children above the median will ultimately benefit everyone *Equal access?* It doesn't happen now *Is fairness more important than beneficence and health?* Beneficence for non-health-related traits *Autonomy*: claiming not a median trait but the parents' choice

Adapted from J. H. Evans' book *The Human Gene Editing Debate*[75]

All the above-mentioned barriers have flaws and one can wonder how long they will hold. Evans also lists a few possible barriers that may arise in the future of the human gene editing debate.

---

[74] Cf. J. H. EVANS, *The Human Gene Editing Debate*, Oxford University Press, New York 2020, 87–92.

[75] Cf. *Ibid.*, 97–108.

8.3.3 Future

See Tables 8.6, 8.7, 8.8, 8.9, and 8.10.

Table 8.6 Goals of medicine Barrier

Goals of medicine barrier
Strength *Medical profession*: beneficence as healing suffering from disease, interpreted as bodily dysfunction, not treating people's desires, wish-fulfilling medicine
Supporting values The same of the disease-enhancement barrier: *Beneficence:* Above the barrier: beneficence is interpreted as relief from suffering. Below the barrier: no-beneficence is interpreted as egotism by wanting to be better than everyone else *Following God's will:* Above the barrier: repairing diseases that were a result of the Fall. Bellow the barrier: trying to improve on God's design for the species
Weakness *How to define the goals?* Broad definition of health by WHO. And the trajectory of medicine is away from limiting its domain to disease and suffering, and towards satisfying the desires of patients (e.g. by offering enhancements)

Adapted from J. H. Evans' book *The Human Gene Editing Debate*[76]

Table 8.7 Family genes barrier

Family genes barrier
Strength *Only traits the couple could have produced through sexual reproduction*
Supporting values *More conservative version of fairness*: there is no redistribution, "you work with what you have" *Loose interpretation of following the will of God:* interpreted as genetically related to parents
Weakness *Concept of the family*: search for advantages of genetic variants in the extended family gene pool (siblings, cousins, etc.…) *Beneficence*: to a child with traits that the couple could not produce *Autonomy*: to install genetic traits from outside the family

Adapted from J. H. Evans' book *The Human Gene Editing Debate*[77]

[76] Cf. *Ibid.*, 113–115.
[77] Cf. *Ibid.*, 115–122.

Table 8.8 Boundary of humanity barrier

Boundary of humanity barrier
Strength *Insertion of human variants only* *Clear limit*: human genes, no continuity or similarity vagueness *Clear concept*: there is consensus on what life forms are human
Supporting values *Following nature and/or God's image*: Above the barrier: insertion of a variant that any human in history has had. Bellow the barrier: insertion of any gene variant that a human has not had (animal variants, artificial variants). Hybrids as a violation of nature
Weakness *Beneficence and autonomy* would allow using non-human variants *Anthropocentrism as opposed to ecology* *So close to the bottom that nobody would bother to build it*

Adapted from J. H. Evans' book *The Human Gene Editing Debate*[78]

Table 8.9 Liberal eugenics barrier

Liberal eugenics barrier
Strength *Not coercive, but voluntary eugenics*: Parents have the autonomy to engage in any gene editing they desire, *short of harming others*
Supporting values *Autonomy and non-maleficence (at the limit).* Beneficence is not really a supporting value since it is defined through the autonomous viewpoint of the parents, who decide what traits are diseases and what traits are improvements if changed
Weakness *Autonomy and non-maleficence* only would make the extremes fall on *similarity vagueness*, potentially bringing down the barrier *Ideally "liberal"* but the *financial burden* affects autonomy and *social norms* can be quite coercive

Adapted from J. H. Evans' book *The Human Gene Editing Debate*[79]

Table 8.10 Scientific reality barrier

Scientific reality barrier
Strength No need to worry about the acts far downslope because they will always be *impossible or are so many decades in the future* that it is not worth thinking about now
Supporting values *No values*, just facts
Weakness It crumbles with our *advancing knowledge* *Not anchored* in any values

Adapted from J. H. Evans' book *The Human Gene Editing Debate*[80]

---

[78] Cf. *Ibid.*, 122–126.
[79] Cf. *Ibid.*, 126–131.
[80] Cf. *Ibid.*, 148.

The above Tables 8.6, 8.7, 8.8, 8.9, and 8.10 were presented as possible future barriers. But "while there was once consensus about the somatic/germline and disease/enhancement barriers, the debate is now without any agreed upon limits."[81]

8.3.4 Transhumanism Enters the Debate

The gene editing debate is difficult not only because of the associated values, but also because it questions what constitutes the desired top and the feared bottom. Evans added this important factor: "The transhumanists remind us that while I am describing the mainstream public bioethical debate about HGE, there are others who see the debate from a totally opposite perspective."[82] Transhumanists would argue that the desired top is an improved version of the species. "In the earliest years of the debate, the eugenicists would have argued that the entire point of the debate was to climb the slope up to the utopia where humans had transformed themselves into a superior species. The eugenicists still exist in this debate, but now they call themselves 'transhumanists' (…)."[83]

This trend was also noted by the European Group on Ethics of Science and New Technologies, in the document *Values for the Future*. "Just as we started to find some relatively firm foundations for ethics in our common humanity, our evolutionary biology, psychology and common history, some suggest that we consider humanity as something that is not given and can be overcome and transcended by technological design and engineering."[84]

For transhumanists, what is good depends on what we decide human beings could or should become. Transhumanists have listed some of their values, with the *core value* being having the opportunity to explore the transhuman and posthuman realms. The *derivative values* include: "tampering with nature"; individual choice in the use of enhancement technologies, peace and international cooperation; improving understanding (encouraging research and public debate; critical thinking; open-mindedness); getting smarter (individually; collectively; and developing machine intelligence), philosophical fallibility (willingness to reexamine assumptions); pragmatism, engineering and entrepreneur spirit; diversity (species, races, religious creeds, sexual orientations, lifestyles, etc.), caring about the well-being of all sentience; and saving lives (life extension, anti-ageing research, and cryonics).[85]

---

[81] *Ibid.*, 140.
[82] *Ibid.*, 145.
[83] *Ibid.*, 144.
[84] European Group on Ethics in Science and New Technologies, *Values for the Future: The Role of Ethics in European and Global Governance*, Publications Office of the European Union, Brussels 2021, in https://ec.europa.eu/info/files/values-future-role-ethics-european-and-global-governance_en [16-11-2021], 13.
[85] Cf. N. Bostrom, «Transhumanist Values», *Journal of Philosophical Research* 30/9999 (2005), 13.

While some claim that transhumanist proposals will lead humanity towards the dystopian bottom, transhumanists say their proposals will not eliminate humanity but rather express it truly. For instance, Sandberg states: "From my perspective, morphological freedom is not going to eliminate humanity—but to express what is truly human even further."[86]

Considering the transhumanist mentality, a great effort would be needed to define a common vision of the desired top and the unwanted bottom in the HGE debate. Will the public agree that the top is an improved humanity? Will there be a common understanding of what constitutes an improvement? What about the unwanted bottom? Will there be agreement on what is not desirable for humanity?

In the middle of many foreseeable disagreements, it is valuable to note there seem to be a few convergences. Bostrom lists some common agreements in *A History of Transhumanist Thought*. For instance, both those against and for enhancements agree that:

- Murder and enslavement, whether of humans by posthumans or the other way around, would be a moral atrocity and a crime.
- Racist and coercive state-sponsored eugenics programs of the twentieth century should not be repeated.
- We face a realistic prospect that technology could be used to substantially transform the human condition in this century.
- This imposes an obligation on the current generation to think hard about the practical and ethical implications.
- There are concerns regarding the medical risks of side effects.
- Technology and medicine have a legitimate role to play.[87]

The first two constitute the common bottom that both transhumanists and non-transhumanists do not want to reach. There seems to be a shared fear of reaching a dystopian bottom. No one wants murder, slavery, racism, and coercion to be part of humanity's future. Both against-enhancement and pro-enhancement scholars reject the scenario presented in Huxley's *Brave New World*.[88] However, against-enhancement scholars blame enhancement for disaster, and pro enhancement scholars blame coercion.

The group of against-enhancement scholars think *Brave New World* has succeeded in eliminating disease, war, anxiety, suffering, guilt, and grief. But "at the heavy price of homogenization, mediocrity, trivial pursuits, shallow attachments,

[86] Cf. A. SANDBERG, *Morphological Freedom—Why We not just Want it, but Need it*, in (2001), in http://www.aleph.se/Nada/Texts/MorphologicalFreedom.htm [16-12-2020]

[87] Cf. N. BOSTROM, «A History of Transhumanist Thought», in M. RECTENWALD - L. CARL (edd.), *Academic Writing Across the Disciplines*, Pearson Longman, New York 2011, 25.

[88] Cf. A. HUXLEY, *Brave New World*, Chatto & Windus, London 1932.

debased tastes, spurious contentment, and souls without loves or longings."[89] The *Brave New World* has achieved prosperity "only to be peopled by creatures of human shape but stunted humanity. (…) Art and science, virtue and religion, family and friendship are all passe. What matters most is bodily health and immediate gratification."[90] In contrast, transhumanists argue that "the best way to avoid a Brave New World is by vigorously defending morphological and reproductive freedoms against any would-be world controllers."[91]

To conclude this section, there seems to be no shared vision regarding the desirable top for humanity, but there are some commonalities regarding the dystopian bottom. There is a difficult horizon ahead for upholding the same values and a shared interpretation of these values.

Evans explores whether any barriers can hold, and his answer is quite disappointing. "Even barriers that were not susceptible to slippery slope processes seemed particularly threatened by powerful values in this debate, such as beneficence."[92] Furthermore, Evans added that the peer review and feedback confirmed his intuition that none of the listed barriers were sufficiently strong. "In the presentations I have made on this project, and in the informal review of the manuscript by colleagues, no one identified strengths I had ignored—only more reasons why a barrier could not hold."[93] There is one additional pressure associated with any given barrier: even if a country creates barriers on the slope, another country will not, and soon enough people will challenge national legislation. Evans concluded that "while I originally thought I was carefully describing how people in the debate could defend a barrier, I may be carefully describing how each barrier will eventually fall in turn. Time will tell."[94]

8.4 Whether, Why, and How to Participate in the Dialogue

What if no ethical barrier can hold strong? Evans recommend defending barriers and forwarding values: "As daunting as defending barriers appears to be, it may be the easier way to forward your values."[95] It is hard to agree with that conclusion, as no promising barrier was found. It sounds as though barriers should be defended not because they are convincing, but because they can protect certain values for a

[89] L. KASS, «Preventing a Brave New World», in *Ethics and Emerging Technologies*, Palgrave Religion & Philosophy Collection, Palgrave Macmillan, London 2014, 76–89, in https://web.stanford.edu/~mvr2j/sfsu09/extra/Kass3.pdf [7-3-2020].
[90] *Ibid.*
[91] N. BOSTROM, «In Defense of Posthuman Dignity», *Bioethics* 19/3 (2005), in https://onlinelibrary.wiley.com/doi/abs/10.1111/j.1467-8519.2005.00437.x [10-4-2021], 206.
[92] J. H. EVANS, *The Human Gene Editing Debate*, Oxford University Press, New York 2020, 140.
[93] *Ibid.*
[94] *Ibid.*
[95] *Ibid.*, 142.

little while longer. In contrast, some thinkers say it would be better to encourage the public to prepare for a world in which enhancements and heritable changes are a reality. Foreseen this scenario, the public should engage in discussing the rights and protections that the various inhabitants of such a world will need.[96]

In this technological and plural society, is it worth trying to join the HGE debate? Is it even worth making proposals from a specific moral framework? The flipped question should also be asked too: what could happen if we leave the dialogue? There are many ways to leave a public dialogue. Some might propose ethical guidelines that can be followed only by groups that follow the same moral standards. That would promote the isolation of minorities, and in practice, it is also a way of leaving the public dialogue. This isolation could have bad consequences for policymakers, doctors, patients, consumers, civil authorities, and society.

There are various reasons to participate in the public dialogue on HGE:

- *There is always an opportunity to do good, to take a step towards good.* As long as human history lasts and man has free will, there is no situation so dark that it cannot be changed. Even when all dangers cannot be avoided, there are always opportunities to prevent the worsening of a situation. Doing some good is always possible.
- *There are encouraging examples of changing past practices to more ethical ones.* Not every aspect of reality has experienced a moral downgrade. There are some encouraging examples where there has been a move towards more ethical living. For example, slavery, racism, and misogyny were socially accepted at certain periods in history and are now repudiated. Female genital mutilation has been condemned at the international level. Surrogacy in India has been disallowed to foreigners, in order to reduce cases of exploitation. Anonymous gamete donation has been banned in the United Kingdom, Canada, Australia, and various European countries.
- *We are social beings, called to contribute to the common good.* Creating bubbles with people who think alike will not solve the problem. Furthermore, sooner or later the problem will turn back. We need to think as social and communitarian beings. If we think more about how to contribute, rather than about asserting our own ideas, it will likely be easier to engage in debate.
- *At least formally, there is a willingness to listen to different voices.* In current times, values such as democracy, diversity, and minorities are important. Therefore, those who *can* but *do not wish to* participate are losing the opportunity to listen to other perspectives and to have their perspectives listened to. Participation is free, but considering the opportunity, there is also a responsibility. People and groups that leave the current dialogue should realise that their absence contributes to creating the future that they do not want. Perhaps the future will become a liberal eugenics society, with the only respected value being autonomy interpreted as individual choice. Decision-makers might rightfully consider

[96] Cf. Eric Juengst in *Ibid.*, 141.

they heard from everyone who *wanted* to be heard. Where were the people with values other than autonomy? Where were the people with different interpretations of autonomy? Simply not there for public debate, because they had lost faith in the dialogue or it seemed too hard?

For those with the will to participate, what can be done? How can they participate? Presented below are some ideas regarding how to participate in the debate of HGE:

- *Depart from the current situation.* We should work with what we currently have, rather than lamenting it is not ideal. This implies accepting the age we live in. Nostalgia for a "better" or "less complicated" past might block action, creativity, and responsiveness. One can look at emerging difficulties but one should also look at what can be done.
- *Discern the truth in the "moral stranger's" vision.* American philosopher Tristram Engelhardt introduced the expressions "moral friends" and "moral strangers". "Moral friends" are those with whom we share moral agreement, alongside rules of evidence and inference. On the other hand, "moral strangers" are those with whom we do not share ethical premises, and this conflict may be due to an incompatible way of reasoning or discrepancy in moral values.[97] Our post-modern, globalised, and pluralistic society is composed not only of "moral friends" but also of "moral strangers". How can we avoid a confrontation that leads to conflict and destruction? How can we live in a society composed of "moral strangers" without falling into a relativism that no longer seeks the common good? According to Austriaco, it is necessary not only to give the "moral strangers" the benefit of the doubt but also to try to discern the truth within their moral vision.[98]
- *Use an effective model for dialogue.* One possible model to participate in the public dialogue is the Tradition-constituted inquiry.[99] This model consists of demonstrating the truth claims of the given tradition by using the epistemological standards of moral strangers. It implies first presupposing the validity of all the opponent's assumptions and their particular standards for rational justification. It implies first presupposing the validity of all the opponent's assumptions and particular standards for rational justification. And then, showing either that they would still lead to inconsistencies and self-contradiction in his tradition,

[97] Cf. H. T. ENGELHARDT, «Morality, Universality, and Particularity: Rethinking the Role of Community in the Foundations of Bioethics», in J. T. L. PO- WAH (ed.), *Cross-Cultural Perspectives on the (Im)Possibility of Global Bioethics*, Philosophy of Medicine, Springer, Dordrecht 2002, 19–38, in https://doi.org/10.1007/978-94-017-1195-1_2 [29-12-2022].

[98] Cf. N. P. G. AUSTRIACO, *Biomedicine and Beatitude: An Introduction to Catholic Bioethics*, The Catholic University of America Press, Washington, D.C. 2021 (2), 388–389.

[99] To see the exposition of this model by Scottish-American philosopher MacIntyre, Cf. A. MACINTYRE, *Whose Justice? Which Rationality?*, University of Notre Dame Press, Notre Dame 1988, 349–369.

or that they would be affirmed by his worldview.[100] This effort can be compared to communicating with someone who speaks a different language. At least one person must learn to speak another language to engage in dialogue with the other. Learning another language does not make the multilingual forget his mother tongue. As Austriaco says, the effective multi-tradition translator must seek to be a virtuous person. The moral translation effort takes not only much patience but also great charity.[101]

- *Aim at the common good.* Sometimes it is necessary to go beyond analysing a simple comparison of views or a critique of a framework. Public good implies the recognition that, on some occasions, we cannot continue in opposite ways. When making decisions that will have consequences for society, it is necessary to engage in dialogue aiming not at mutual understanding only, but also at solutions to build a healthy community.
- *Find your personal specific way(s) to participate in the public dialogue.* An individual's participation can be translated into different actions, depending on their interests, background, profession, and specific opportunities that they have access to. Forms of engagement can range from having personal conversations, to organising or participating in local/national/international debates and conferences, responding to surveys/public referendums, participating in related advocacy groups, becoming involved in legal decisions/policies/allocation of funding/research efforts, producing scientific publications, producing media publications, teaching, giving counselling to individuals or groups, and so on.

[100] Cf. N. P. G. Austriaco, *Biomedicine and Beatitude: An Introduction to Catholic Bioethics*, The Catholic University of America Press, Washington, D.C. 2021 (2), 376.
[101] Cf. *Ibid.*, 379.

Refining the Predominant Values for the Global Governance of Human Genome Editing

9

Abstract

A common theme that resulted from the Reflexive Thematic Analysis (RTA) method applied to selected WHO and Humanity+ texts is *global governance to minimise the risks and maximise the benefits of human genome editing*. This theme comprises the main governance elements and reveals some shared values. These values were categorised into three: respect for individual rights (interpreted as autonomy in health and reproduction), protection of people who cannot express their wishes (referring mainly to future generations), and equality-solidarity. Current Bioethics has the task of refining these predominant values. This refinement is done by reinterpreting or complementing the meaning of these three given values. So, *respect for individual rights-autonomy* ceases to mean only the fulfilment of personal wishes in health and reproduction and connects to other values such as humility, responsibility, and sociability, also attentive to explicit and implicit coercions. The *protection of people who cannot express their wishes*, a value which refers mostly to future generations, is broadened to the protection of all the vulnerable. As for the value of *equality and solidarity*, it gains a consideration about intrinsic equality and extrinsic diversity, to be solid and avoid inconsistency.

Keywords

Governance • Values • Bioethics • Human genome editing • Public debate

L. Santos, *Governance of Human Gene Editing and Transhumanism*,
Integrated Science 28, https://doi.org/10.1007/978-3-031-70863-3_9

9.1 The Meaning of Refining Values

A common theme that resulted from the Reflexive Thematic Analysis (RTA) method applied to selected WHO and Humanity+ texts is *global governance to minimise the risks and maximise the benefits of human genome editing.*[1] This theme comprises the main governance elements and reveals some shared values. These values were categorised into three: respect for individual rights (interpreted as autonomy in health and reproduction), protection of people who cannot express their wishes (referring mainly to future generations), and equality-solidarity. Current Bioethics has the task of refining these predominant values. Refining is the process of removing impurities or unwanted elements from a substance. For example, oil is refined into gasoline to provide fuel. Iron is refined to make steel for construction. The crude gold is melted and then treated with chloride. The process converts any foreign elements still in the gold into chloride, which then naturally drifts off the gold. Refining is not replacing. The given resource is good, but refining takes place so natural resources can achieve a higher value or attain a more effective usage. Analogously, the predominant values identified by the RTA method can be refined. There is something precious in all the common values present in the governance of HGE and Transhumanism. Nonetheless, they could be more effective if they are refined.

Evans's remarkable effort in the book *The Human Gene Editing Debate* was centred on analysing the barriers set to avoid slippery slopes in the use of HGE. He explains the values supporting each barrier and their weaknesses.[2] Barriers are the negative part, the "do not cross this line". But what if the effort is also placed on enlightening, refining, and complementing the predominant values in the HGE debate? No value is an isolated entity, values are rather connected. Understanding any value as an isolated one misleads ethical reasoning, as values can easily enter into conflict, and it may seem that upholding a given value will inevitably harm others. This conflict is brilliantly shown in Evans' analysis of the weakness associated with the barriers in the HGE debate. Much of the problem seems to be the consideration of values as isolated entities, fostering conflict between values, or implying a reductionist understanding of a specific value. Moving towards interconnected values would create a net of values. Perhaps a net of values will end up being stronger than a barrier supported by values.

[1] The full Reflexive Thematic Analysis (RTA) can be consulted in the published doctoral dissertation excerpt L. SANTOS, *Common Themes Between Humanity+ Original Documents on Transhumanism and WHO Documents on the Global Governance of Human Genome Editing,* Ateneo Pontificio Regina Apostolorum, Rome 2023.

[2] Cf. J. H. EVANS, *The Human Gene Editing Debate*, Oxford University Press, New York 2020.

9.2 Refining "Individual Rights-Autonomy in Health and Reproduction"

Regarding the first value, the results of the RTA showed that *respect for individual rights* is an important aspect seen within Humanity+ and WHO texts. The expression "individual rights" was kept with respect to the texts. But, as previously explained, "individual rights" refers especially to *autonomy,* meaning autonomy of individual choice, and is almost always regarding health and reproduction.

Autonomy is indeed an important value. The opposite would be coercion, which is highly repulsive. Coercion in its various forms such as forced labour, forced marriage, lack of informed consent, and other forms of unfree decisions in important aspects of the individual's life is not desired. Lack of autonomy in these situations means the person being coerced is not considered as intelligent, capable of choices, or a free human being. Diverse forms of coercion have brought devastating consequences to many people throughout human history.

However, we should ask: Why does autonomy today seem to be above many other possible values? For instance, the other two values found in the RTA applied to WHO and H+ texts are also closely connected to autonomy. The value of protection refers to the protection of people who cannot express their wishes. What is being valued is the liberty of people who cannot communicate their autonomy, either because they are not born yet (in the case of future generations), or because they lack the actual capacity to exercise their autonomy (e.g., legal minors, people with certain disabilities, etc.).

The other value found in the RTA refers to equality-solidarity. This one seems to be more closely related to justice. However, if autonomy is the paramount value that sets the tone, what will be considered as fair? Autonomy and individual views influence equality, and in practice, autonomy prevails over fairness. I will address these other two values- protection and equality- later on in this work.

For now, the point is that there seems to be an implicit predominance of the value of autonomy. However, should this be the case? It may be the current scenario in public debate, but is it the best possible scenario? Some have argued that "the transhumanist position does not sufficiently justify its arguments, nor does it develop a solid rationale to explain why the concepts of autonomy and self-transformation should be more valuable than any other values (even within the techno-scientific framework)."[3] What might happen if autonomy, interpreted as individual choice regarding health and reproduction, is the leading value? What is the foreseeable scenario?

Whether isolated or at the top of the hierarchy, the value of autonomy can mess up a healthy life in society and the life of the individual as well. Since biology is no longer considered a limit to autonomy, claims for freedom of form

[3] J. Piedra, «Technological Enhancement and Happiness: A Review of Morphological Freedom», *Cosmos and History: The Journal of Natural and Social Philosophy* 15/1 (2019), 275–291, in http://www.cosmosandhistory.org/index.php/journal/article/view/805 [13-4-2021], 286.

emerge. As written in the book *From Transgender to Transhuman,* "Freedom of gender is, therefore, the gateway to a freedom of form and to an explosion of human potential. First comes the realization that we are not limited by our sexual anatomy. Next comes the awakening that we are not limited by our anatomy at all."[4] Anatomy is not considered a boundary to liberty, moreover, it is becoming the object of liberty. As Sandberg writes in *Morphological Freedom—Why We not just Want it, but Need it*: "From the right to freedom and the right to one's own body follows that one has a right to modify one's body. If my pursuit of happiness requires a bodily change—be it dying my hair or changing my sex—then my right to freedom requires a right to morphological freedom."[5] Sandberg affirms that physical welfare may require affecting the body and explains that "morphological freedom is the right to modify oneself. (…) We express ourselves through what we transform ourselves into."[6]

If autonomy is the determinant of morality, it will define what is good for reproduction, health, and enhancements. This scenario may even kill the idea of enhancement. Autonomy will likely go beyond liberty as to whether to enhance or not. Individual autonomy will likely decide what constitutes an enhancement or a dis-enhancement. The meaning of enhancement may move from betterment towards a self-defined matter. Then, advocacy for enhancement may be perceived as interference with the limitless individual autonomy to define what is good and better. If the point is self-expression, discussions regarding normalcy, enhancements, or dis-enhancements will be left behind.

Perhaps instead of the word *enhancement*, which involves a recognition of what is *good* and what is *better*, the word *modification* expresses more accurately what is being proposed by transhumanists and what liberal societies are heading towards. The controversy will likely switch from the possibility of *enhanced,* to the possibility of freely *modified* people.[7] In this scenario, liberty as self-expression, with deeper consequences in the body, might be the only value protected by law. In Sandberg's words: "Taking the step to full morphological freedom creates a far simpler ethical guideline, which both protects those who do not wish to change, those who are differently bodied and those wanting to change their bodies."[8] But

[4] Cf. M. Rothblatt, *From Transgender to Transhuman: A Manifesto On the Freedom Of Form*, Martine Rothblatt, USA 2011, 29.

[5] Cf. A. Sandberg, *Morphological Freedom – Why We not just Want it, but Need it*, in The Transhumanist Reader, John Wiley & Sons, Ltd, Chichester 2013, in https://doi.org/10.1002/9781118555927.ch5 [14-4-2020], 56–64.

[6] Cf. *Ibid.*

[7] See more about modifications and concrete examples in Cf. C. H. Gray - H. Figueroa-Sarriera - S. Mentor, *Modified: Living as a Cyborg*, Routledge, New York 2020, in https://doi.org/10.4324/9781351107839 [12-12-2020]. And also Cf. K. Tsui, «Transhumanism: Meet the cyborgs and biohackers redefining beauty», *CNN* (2020), in https://www.cnn.com/style/article/david-vintiner-transhumanism/index.html [22-3-2021].

[8] Cf. A. Sandberg, *Morphological Freedom – Why We not just Want it, but Need it*, in The Transhumanist Reader, John Wiley Sons Ltd, Chichester 2013, in https://doi.org/10.1002/9781118555927.ch5 [14-4-2020], 56–64.

what might happen? Every attempt at regulation would be pointless, as the line would be constantly challenged due to new technological advances. And, more importantly, rules would be perceived as arbitrary and end up not protecting people in need of protection.

9.2.1 Reconnecting Autonomy

As earlier stated, autonomy is to be respected and given significant consideration. The problem is not the value, but the comprehension of autonomy in an isolated way and placing it above all other values. Autonomy is an aspect of being human that must be considered comprehensively to be understood. Alone, it does not capture the completeness of human beings. Autonomy alone does not reflect "the relationship between the cognitive, social and physical-corporal aspects"[9] of human life, and forgetting these aspects has consequences.

By removing the considerations about human nature, the human condition, the relationship between the body and the psyche, and the social aspect of humans, there is space for everything, good and bad. Many have pointed out that the main problem of current Bioethics is the lack of reference to human nature. "Without it, ethics is reduced to a series of private experiences disconnected from responsibilities that have to do with others and with society."[10] In this scenario, "each individual attends to their own moral standards or values and uses them for their own particular good, without caring about the problems and well-being of others who surround them, as long as they respect their rights in a multi-moral context."[11]

Although the liberal paradigm creates problems by not using nature as a reference, it is difficult to specify what human nature is. There were insufficient attempts to discuss human nature in a multi-moral context. For instance, Fukuyama's vague "factor X"[12] in his attempt to specify the essential human quality which makes all humans worthy.

A more valuable attempt was made by Snead in his book *What It Means to Be Human: The Case for the Body in Public Bioethics*. "The truth is that persons are *embodied* beings, with all the natural limits and great gifts this entails. We experience our world, ourselves, and one another as living (and dying) bodies."[13]

[9] J. PIEDRA, «Technological Enhancement and Happiness: A Review of Morphological Freedom», *Cosmos and History: The Journal of Natural and Social Philosophy* 15/1 (2019), 275–291, in http://www.cosmosandhistory.org/index.php/journal/article/view/805 [13-4-2021], 286.

[10] L. M. PICHARDO, «Reinventing Bioethics in a Post-humanist and Post-truth Society. The Present and Future of Bioethics», *Persona y Bioética* 22/2 (2018), 212–222, in https://doi.org/10.5294/pebi.2018.22.2.2 [15-8-2020], 219.

[11] *Ibid.*

[12] Cf. F. FUKUYAMA, *Our Posthuman Future: Consequences of the Biotechnology Revolution*, Profile Books, New York 2003.

[13] O. C. SNEAD, *What It Means to Be Human: The Case for the Body in Public Bioethics*, Harvard University Press, Cambridge 2020, 3.

Snead begins by exploring our experience of being embodied beings and then identifies what this human embodiment implies. Our own bodies talk about being vulnerable and mutually dependent. "Because we are bodies, vulnerability, mutual dependence, and natural limits are inextricable features of our lived human reality."[14] The embodiment fact "situates us in a particular relationship to one another, from which emerge obligations to come to the aid of vulnerable others, including especially the disabled, the elderly, and children."[15]

Other authors, such as Max Scheler (German philosopher, 1874–1928)[16] and Arnold Gehlen (German philosopher, sociologist, and anthropologist, 1904–1976)[17], have identified the implications of the human embodied condition to anthropology. As noted by Prieto López, in the 20th century an interesting and promising change of direction occurred in anthropology. *Biological anthropology* approaches the human being from a perspective that, although philosophical, has a new starting point: the human body.[18]

There is more to our humanity than intelligence, freedom, and self-expression. Not only there is more, but even intelligence, freedom and self-expression are incarnated, and connected to our bodily existence and experience. Therefore, the value of autonomy should be combined with other values that reflect our bodily human condition.

In the gene editing debate, "(…) the two experts may be duelling bioethicists, one saying that to allow the experiment will violate the end of nonmaleficence, because it might not be safe, the other saying that to stop the experiment would violate the end of autonomy of parents to engineer their children as they see fit."[19] The rationality of the classical four bioethical principles presented by Beauchamp and Childress (*beneficence*, *nonmaleficence*, *autonomy*, and *justice*)[20] is insufficient to manage human genome editing issues. These four bioethical principles are good but have no hierarchy, so in liberal and individualistic societies ethical reasoning frequently collapses under the predominance of autonomy. Furthermore, some scholars have even rejected the four principles framework because autonomy does not occupy the theoretical central position in Beauchamp and Childress' proposal. For example, Reis-Dennis argues that "respect for autonomy is, and ought to be,

[14] *Ibid.*

[15] *Ibid.*

[16] Cf. M. SCHELER, *The Human Place in the Cosmos*, Northwestern University Press, Evanston 2009.

[17] Cf. A. GEHLEN, *Studien zur Anthropologie und Soziologie*, Soziologische Texte 17, Luchterhand Verlag, Neuwied 1963.

[18] Cf. L. PRIETOLÓPEZ, *El hombre y el animal: Nuevas fronteras de la antropología*, Estudios y Ensayos. BAC. Filosofía y Ciencias, Biblioteca Autores Cristianos, Madrid 2008.

[19] J. H. EVANS, *Playing God? Human Genetic Engineering and the Rationalization of Public Bioethical Debate*, University of Chicago Press, Chicago 2002, 204.

[20] Cf. T. L. BEAUCHAMP - J. F. CHILDRESS, *Principles of Biomedical Ethics*, Oxford University Press, New York 2001 (5).

the fundamental value of bioethics"[21], calling human dignity the "founding myth of Bioethics".[22] In practice, the principle of autonomy tends to be contrasted with the other principles, instead of being completed by them. Using this framework, the tendency is to identify conflicts and name concerns, rather than concluding the ethical reasoning. However, the ladder would be far more useful for society.

Other bioethicists, following the Personalistic framework, use a more comprehensive principle: responsible liberty.[23] In this framework, liberty and responsibility are used together to achieve individual and common good. Responsible liberty is also connected to other values that are proposed as principles. Furthermore, the predominance is given to the value of life, so no interpretation of liberty or any other value can go against it. The principles of Personalist Bioethics are *defence of physical life* (one cannot exist, be free, or have human rights without life), *liberty and responsibility* (to achieve individual and collective good), *totality and therapy* (the organism is organically unified, it is licit to intervene on a part to save the totality of the body), *sociability and subsidiarity* (life is a social and not only an individual patrimony, society must help without substituting individual initiatives. This principle also implies solidarity).[24]

To conclude this section, if autonomy is not connected to other values, it does not match our experienced bodily existence. Also, this value loses useful moderators. Some good moderators of autonomy include the values of *humility* (which recognises that individuals can be mistaken), *responsibility* (which reminds that individuals are accountable for their choices), and *sociability* (which adds that some individual acts have social effects).

9.2.2 Beware of Explicit and Implicit Coercion

Agreeing that autonomy must be respected and coercion is not desired, one can question whether liberty will be truly granted. Liberty needs to be protected from explicit coercion but also from implicit forms of it. In order to promote real autonomy, it is necessary to recognise and diminish explicit and implicit coercion. Explicit coercions are easier to identify, but they are not the only threats to our liberty.

One of the families that participated in Dr. He's clinical trial was asked about possible coercion by the scientists. The family wrote a letter expressing how they felt coerced. They revealed how they felt coerced not by the researchers, but by society. The following is a translation of their words:

---

[21] S. Reis- Dennis, «Dignity and the Founding Myth of Bioethics», *Hastings Center Report* 53/2 (2023), 26–35, in https://onlinelibrary.wiley.com/doi/abs/10.1002/hast.1471 [10-5-2023], 26.
[22] *Ibid.*
[23] Cf. E. Sgreccia, *Manuale di bioetica. Fondamenti ed etica biomedica*, I, Vita e Pensiero, Milano 1994^2, 175.
[24] Cf. *Ibid.*, 171–180.

> In selecting this type of experiment, we were never misled. It was a form of compromise. The object of the compromise was society, and one could even say with the whole world. As an AIDS sufferer and family member, we firmly and deeply know that it is possible to use a preventative drug to have a healthy child... That drug can cure disease, but it cannot cure prejudice... For everyone who is listening, please hear this. At a certain level our participation in the experiment was indeed forced, but we were not coerced by any person in particular. We were coerced by society. (Letter from —, translated from the Chinese, 2019)[25]

It is important to note that autonomous expression of choices does not guarantee liberty. Liberty can be very diminished, even when someone clearly enunciates personal preferences. Implicit coercion, which takes the form of mostly economic and social pressure, has an impact on an individual's "free" choice. For instance, "a decision that produces a financial burden is not truly autonomous, and social norms can be quite coercive."[26] For example, advertisements have always influenced individual desires. Nowadays, this happens exponentially in social media, and society seems to accept referring to certain people as "influencers". Another example of a form of limiting "free choice" is the consent that users often give to website cookies. There are policies in place to protect user privacy, but people quickly consent to sharing personal data to get what is being offered. Another example is that some athletes calculate how to use substances to improve performance, without failing doping tests. Why? Because they cannot afford to be left behind, compared to other competitors. Furthermore, in reproduction today, there seems to be more autonomy because there are more choices. However, a wider range of choices also affects real autonomy.

> Parents who practice sex selection today may be pressured tomorrow to use enhancement technologies to extend a disease-free legacy. Societal and peer pressures might reach a stage where a child born with disabilities will be considered a "wrongful life" with "wrongful disability," and his misguided parents punished accordingly. Extreme freedom of choice can even result in communitarian eugenics, whereby groups choose certain traits in their offspring—deafness, religiosity, homosexuality—to perpetuate the existence of such groups.[27]

Similarly to what happened with other inventions and technologies, genome editing will probably not be imposed on us, but rather be sold as something we cannot live without.[28] With the awareness that, in many other contexts, social pressure implicitly but truly threatens liberty, what will be done to protect liberty in an HGE era?

---

[25] J. B. Hurlbut, «Imperatives of Governance: Human Genome Editing and the Problem of Progress», *Perspectives in Biology and Medicine* 63/1 (2020), 177–194, in https://muse.jhu.edu/article/748059 [24-5-2022], 191.

[26] J. H. Evans, *The Human Gene Editing Debate*, Oxford University Press, New York 2020, 129.

[27] J. Tham, «Resisting the Temptation of Perfection», *The National Catholic Bioethics Quarterly* 17/1 (2017), 51–62, 59.

[28] Cf. Mark Frankel in F. Baylis, *Altered Inheritance: CRISPR and the Ethics of Human Genome Editing*, Harvard University Press, Cambridge 2019, 73.

Bioethics should advocate for the protection of true liberty. Firstly, by connecting autonomy to other values such as humility, responsibility, and sociability. This would mean reframing liberty: instead of emphasising autonomy of choices, moving towards humble, responsible, and socially-aware liberty. The common effort should work towards including other values and networking them, instead of watching autonomy prevail over all other goods. Secondly, the promotion of autonomy implies protection from explicit and implicit coercion. Instead of merely advocating against explicit coercion, common efforts should be made to debunk implicit coercions.

9.3 Refining "Protection of People Who Cannot Express Their Wishes"

Another value that resulted from the RTA of the selected texts is the protection of people who cannot express their wishes. As it was noted earlier, what is being protected is precisely the liberty of people who cannot express their autonomy, either because they are not born yet (the unborn and the case of future generations), or because they lack the actual capacity to exercise their autonomy (minors, some people with specific disabilities, those experiencing limiting diseases, etc.).

The protection of people who cannot express themselves is important. It shows that the concerns surrounding HGE is not centred only on those who can express themselves. This value shows that what is highly desirable for the individual, should also be desired for all other human beings. It shows altruism and solidarity. It recognises that people who cannot express themselves need and deserve something from the people who can express themselves. What do they deserve? Effective protection and defence from harm.

Concerning the protection of future generations, the main issue is heritable human genome editing (HHGE). Doudna had said CRISPR would be used in this way: "And, have no doubt, this technology will-someday, somewhere- be used to change the genome of our own species in ways that are heritable, forever altering the genetic composition of humankind."[29] Only a few months later, this "someday" arrived and this "somewhere" took place, with Dr. He's 2018 experiment in China.[30]

Today, deciding to proceed with HHGE would mean abandoning the restraint urged by the laws and regulations of at least 75 nations,[31] including 29 that have ratified the binding international treaty known as the *Oviedo Convention*, which

[29] J. A. DOUDNA - S. H. STERNBERG, *A Crack in Creation: Gene Editing and the Unthinkable Power to Control Evolution*, Houghton Mifflin Harcourt, Boston 2017, xvi.

[30] Cf. M. MARCHIONE, «Chinese researcher claims first gene-edited babies», AP NEWS (2018), in https://apnews.com/article/4997bb7aa36c45449b488e19ac83e86d [19-12-2020].

[31] Cf. F. BAYLIS ET AL., «Human Germline and Heritable Genome Editing: The Global Policy Landscape», *The CRISPR Journal* 3/5 (2020), 365–377, in https://www.liebertpub.com/doi/10.1089/crispr.2020.0082 [14-7-2021], 369.

states in art. 13: "An intervention seeking to modify the human genome may only be undertaken for preventive, diagnostic or therapeutic purposes and only if its aim is not to introduce any modification in the genome of any descendants."[32] As the Geneva Statement on HHGE affirms, "Policymakers put these prohibitions in place to protect human rights and the fundamental equality of all people; to safeguard the physical, psychological, and social wellbeing of children; and to avert the emergence of a new eugenics."[33]

Even countries that have no regulation around HHGE are strongly discouraged from attempting it due to the moratorium by WHO. The WHO expert advisory committee on HGE convened on 18–19 March 2019 and made an interim recommendation to the WHO Director-general. WHO Director-General Dr Tedros Adhanom Ghebreyesus stated that "it would be irresponsible at this time for anyone to proceed with clinical applications of human germline genome editing."[34] WHO Director-General added that HHGE "poses unique and unprecedented ethical and technical challenges. (…) I have accepted the interim recommendations of WHO's Expert Advisory Committee that regulatory authorities in all countries should not allow any further work in this area until its implications have been properly considered."[35]

A moratorium is not a ban, it is a prohibition at the current time due to certain factors. If conditions are met, the moratorium can be lifted. As Mills notes, "Of course, any statement that includes the phrase 'at this time' need not necessarily apply tomorrow."[36] What has to change for the judgement on HHGE to be modified? Basically, three things: "well-evidenced technical refinement (on which the International Commission focussed), the institution of effective governance (for which the WHO provides a framework and heuristic), and societal and ethical resolve."[37] There is work in progress in the first two areas (technical safety and a framework for global governance), but the resolution of the third condition is difficult to measure. In all three- technical safety, governance, and public debate- the value of protection of future generations plays an important role.

[32] Council of Europe, «Convention for the Protection of Human Rights and Dignity of the Human Being with regard to the Application of Biology and Medicine: Convention on Human Rights and Biomedicine», *European Treaty Series* 1997, in https://rm.coe.int/CoERMPublicCommonSearchServices/DisplayDCTMContent?documentId=090000168007cf98 [15-7-2021], 13.

[33] R. Andorno et al., «Geneva Statement on Heritable Human Genome Editing: The Need for Course Cor-rection», *Trends in Biotechnology* 38/4 (2020), in https://www.sciencedirect.com/science/article/pii/S0167779919303178 [4-3-2021], 352.

[34] WHO, «Statement on governance and oversight of human genome editing», 2019, in https://www.who.int/news/item/26-07-2019-statement-on-governance-and-oversight-of-human-genome-editing [13-7-2021].

[35] *Ibid.*

[36] P. Mills, «Human genome editing and moral leadership: findings of the WHO Expert Advisory Committee», 2021, *The Nuffield Council on Bioethics*, in https://www.nuffieldbioethics.org/blog/human-genome-editing-and-moral-leadership-findings-of-the-who-expert-advisory-committee [14-7-2021].

[37] *Ibid.*

Is HHGE contrary to the protection of future generations? If that is the case today, will that always be the case? Can HHGE be done while protecting future generations? Can HHGE be a means to protect future generations from suffering and diseases? Considering the transhumanist approach, will protection be interpreted and aligned with the principle of Procreative Beneficence (PB)?

Authors like Savulescu claim a procreative freedom/duty: "couples who decide to have a child have a significant moral reason to select the child who, given his or her genetic endowment, can be expected to enjoy the most well-being."[38] And to defend critiques from disability groups advocates they say: "Although PB instructs us to aim to reduce disability in future children whenever possible, it does not privilege the normal. What matters is not whether future children meet certain biological or statistical norms, but what level of well-being they can be expected to have."[39] When Savulescu and his former thesis adviser, Peter Singer, proposed a hierarchy for a translational HHGE, it was also susceptible to critiques from disability groups. Singer and Savulescu made their differentiation by mentioning catastrophic disorders, serious disorders, and common diseases: "In our view, to be ethically justifiable, such a 'translational pathway' should be: catastrophic single gene disorders (like Tay–Sachs disease), then severe single gene disorders (like Huntington's disease), then reduction in the genetic contribution to common diseases (like diabetes and cardiovascular disease), then enhanced immunity and perhaps even delaying ageing."[40]

However, the notion of "serious" genetic disease, which has been broadly used within the legal and policy spheres in the context of HHGE, remains undefined. Some people argue that the burden of serious diseases can be experienced from multiple angles (medical, emotional, economic, and so on).[41] Additionally, the testimony of people with "bad" genes shows the complexity of the situation. There is the fear that gene editing can reduce human diversity and increase social inequality "by editing out the kinds of people that medical science, and the society it has shaped, categorize as diseased or genetically contaminated—people like us who are understood as having bad genes."[42] The authors add that "we should be reminded that bad genes don't necessarily lead to bad lives, just as good genes don't necessarily lead to good lives."[43]

---

[38] Cf. J. Savulescu - G. Kahane, «The Moral Obligation to Create Children with the Best Chance of the Best Life», *Bioethics* 23/5 (2009), in https://onlinelibrary.wiley.com/doi/abs/10.1111/j.1467-8519.2008.00687.x [12-2-2020], 274.

[39] Cf. *Ibid.*

[40] J. Savulescu - P. Singer, «An ethical pathway for gene editing», *Bioethics* 33/2 (2019), in https://onlinelibrary.wiley.com/doi/abs/10.1111/bioe.12570 [17-8-2020], 221.

[41] Cf. E. Kleiderman - V. Ravitsky - B. M. Knoppers, «The 'serious' factor in germline modification», *Journal of Medical Ethics* 45/8 (2019), 508–513, in https://www.ncbi.nlm.nih.gov/pmc/articles/PMC6820154/ [14-7-2021], 512.

[42] S. S. Garland- Thomson Rosemarie, «The Dark Side of CRISPR», 2021, *Scientific American*, in https://www.scientificamerican.com/article/the-dark-side-of-crispr/ [3-3-2021].

[43] *Ibid.*

Some people with disabilities want their lives to continue as they are, fearing their community might disappear and they speak up. However, some people cannot even express their wishes, and herein lies an important concern regarding HHGE. Transhumanists such as Bostrom and Roache conclude that there are no compelling reasons to resist gene editing to select the best children. "There are, however, important issues relating to the fact that such intervention would involve the selection of traits of a person who has no say in the matter"[44], and that is the reason behind the "importance to consider at all times the best interests and future welfare of the resulting children."[45]

The "best interest of the child" is what protects minors today from mistreatment and negligence. It is the foundation for the currently approved measures that may limit parental freedoms, for instance: laws against child abuse and child neglect, obligatory schooling, and some cases of legally enforced medical treatment given to a child even against the wishes of the parents. In the case of unborn children or of future generations to come, who judges what is in "the best interest of the child"? What is the criteria? Choices are not value-neutral. Traits we value today may not be valuable in the future. For example, "(...) if human genetic engineering had been available in Victorian times, people might have designed their children to be patriotic and pious".[46] Therefore, Bostrom and Roache propose a kind of "objective" enhancement: "characteristics that are likely to benefit the child regardless of her eventual preferences and values, and regardless of her cultural context. Characteristics such as intelligence, happiness, and health are more likely to serve this end than characteristics like piety, competitiveness, and sporting prowess."[47] However, it is also clear they value intelligence, well-being, and health. Similar traits are listed by Bostrom: "Some mothers and fathers might find it easier to love a child who, thanks to enhancements, is bright, beautiful, healthy, and happy."[48]

Heritable gene editing may allow parents to promote certain characteristics in their child. But, even if one day this can be safely done, it will be accompanied by two problems. First, it will likely increase parental expectations, which is a burden for the child and might reduce his freedom.[49] Second, if heritable gene editing becomes an option, "each person, whether she has been genetically programmed or not, can regard her own genome as the consequence of a criticizable action *or*

[44] N. BOSTROM - R. ROACHE, «Ethical Issues in Human Enhancement», in J. RYBERG - T. PETERSEN - C. WOLF (edd.), *New Waves in Applied Ethics, New Waves in Philosophy*, Palgrave Macmillan UK, Hampshire and New York 2007, 149.
[45] *Ibid.*
[46] Jonathan Glover in *Ibid.*, 147.
[47] *Ibid.*
[48] Cf. N. BOSTROM, «Human Genetic Enhancements: A Transhumanist Perspective», *The Journal of Value Inquiry* 37/4 (2003), in https://doi.org/10.1023/B:INQU.0000019037.67783.d5 [12-7-2021], 493–506.
[49] Cf. NUFFIELD COUNCIL ON BIOETHICS, *Genome editing and human reproduction: Social and ethical issues*, Nuffield Council on Bioethics, London 2018, in https://www.nuffieldbioethics.org/publications/genome-editing-and-human-reproduction [20-4-2018], 72.

omission."[50] The modifications will not always result in good for the child, and the parents will be held accountable for their genetic choices. This scenario "raises the question of whether we can ever assume the responsibility for the distribution of natural talents".[51]

9.3.1 Protection of All the Vulnerable

These difficulties arise because the value of protection is centred on protecting people who cannot express their wishes. This is different from the protection of all people in need of protection—the vulnerable. Instances such as euthanasia or voluntary death makes this difference clear. If the vulnerable sick or suicidal person is not protected, they might end their life in a moment of intense suffering and despair. Their liberty will be apparently respected, but their very lives will be left unprotected in this moment of intense vulnerability.

A better refinement of the value of protection would be placing an emphasis on vulnerability. Instead of simply defending people who cannot express themselves, protection should be given to all people in need of protection, which means to all vulnerable people. This includes people who cannot express themselves, the unborn, the future generations, the physically and socially disadvantaged people, and those who are coerced by society or by their own suffering when expressing themselves. A better path forward is to extend the value of protection to all vulnerable people.

9.4 Refining "Equality-Solidarity"

Among the WHO values and principles to inform *what* decisions are made regarding HGE, one value is more related to autonomy and individual rights (*Respect for persons*), with the meaning already explained when addressing the value of autonomy. Another value is more related to protection and safety (*Caution*). Overall, most of the values for decision-making chosen by WHO were related to equality and solidarity: *Inclusiveness, Fairness, Social justice, Non-discrimination, Equal moral worth, Solidarity* and *Global health justice.* Transhumanist texts also expressed concern for inequality. They say the improvement they dream of for humanity should not be like pre-World War II eugenics. "In addition to condemning the coercion involved in such policies, transhumanists strongly reject the racialist and classist assumptions on which they were based (…)"[52] When defending enhancements, transhumanists argue that they should be available for all, and not lead to discrimination.

50 J. HABERMAS, *The Future of Human Nature*, Polity Press, Cambridge 2003, 82–83.
51 *Ibid.*
52 HUMANITY+, «Transhumanist FAQ 3.0», 2001, Humanity+, in https://www.humanityplus.org/transhumanist-faq [2-12-2020].

In the final RTA analysis of WHO and H+ texts, equality and solidarity are presented as two sides of the same coin, encompassing a set of related values (*Inclusiveness, Fairness, Social justice, Non-discrimination, Equal moral worth, Solidarity* and *Global health justice).* This set of values is important and has the potential to moderate autonomy. Equality-solidarity implies recognition that everyone has equal moral worth and deserves to be treated fairly and helped in times of need. Equality asks for non-discrimination and global health justice, while solidarity implies understanding equality to live in fraternity and reciprocity.

This all sounds very nice, but what kind of equality are we talking about? Equality of access? Equality of opportunities? Equality in the outcomes? Intrinsic equality? Extrinsic equality? Equality of access is supposed to be included in health justice, but in practice, equality of access does not happen today. One recent example can be observed in the *Global Dashboard for Vaccine Equity.*[53] In addition to a failing reality, equality of access seems insufficient to ensure the value of equality and solidarity. According to Savulescu and Singer: "Perhaps the best we can hope for is harm reduction and a regulated market to make important enhancements, such as resistance to disease or the enhancement of intelligence (should it ever be possible), part of a basic healthcare plan so that the benefits of gene editing are distributed equally."[54]

Even if all people around the globe had equal access to gene editing, there would still exist socio-economic inequalities. Bill Gates once said: "I'm a huge believer in that science and innovation are going to solve most of the tough problems over time."[55] This was a sort of "In Science we trust" statement. And yet, "technology has not solved many major social issues. People around the world continue to lack access to sanitation, food, education, and health care. It is not a lack of technology that allows these problems to continue, but rather a lack of political will."[56]

Before thinking of equality of access to HGE, it should be acknowledged that inequality starts in the collection of data, which impacts genetic research and the people who will benefit from it. For instance, in the Genome-Wide Association Studies (GWAS) catalogue, 79% of all participants were of European descent, despite making up only 16% of the global population.[57]

[53] Cf. UNDP, «Global Dashboard for Vaccine Equity», 2022, *UNDP Data Futures Platform*, in https://data.undp.org/vaccine-equity/ [30-3-2022].

[54] J. SAVULESCU - P. SINGER, «An ethical pathway for gene editing», Bioethics 33/2 (2019), in https://onlinelibrary.wiley.com/doi/abs/10.1111/bioe.12570 [17-8-2020], 222.

[55] Bill Gates in N. LEE, «Brave New World of Transhumanism», in N. LEE (ed.), *The Transhumanism Handbook*, Springer International Publishing, Cham 2019, 3–48, 14.

[56] A. M. DEBAETS, «Rapture of the Geeks: Singularitarianism, Feminism, and the Yearning for Transcendence», in *Religion and Transhumanism: The Unknown Future of Human Enhancement*, ABC-CLIO, Santa Barbara 2014, 181–197, 190–191.

[57] Cf. A. R. MARTIN ET AL., «Clinical use of current polygenic risk scores may exacerbate health disparities», *Nature Genetics* 51/4 (2019), 584–591 1.

Equality of access and global distribution of HGE technologies were addressed in the 2023 Third International Summit on Human Genome Editing.[58] After the first person cured of sickle-cell disease (SCD) using gene therapy gave her testimony, a speaker who is a representative of an SCD Organization in India pointed out that the ordinary medication for SCD today is not affordable in India, forty years after its development. He questioned whether the same fate is awaiting gene editing therapy for SCD patients in India, one of the countries with the highest number of individuals suffering from SCD.

During the Summit, it was also mentioned that the cost of developing a gene therapy is around $5 billion, which is five times the cost of developing a traditional drug. Furthermore, the cost of gene therapy is about $1 to 3.5 million. Therefore, policymakers can approve a specific gene therapy and insurance companies can decide not to cover it. As Steve Pearson, from the Institute for Clinical and Economic Review, said at the London Summit, SCD gene therapy is expected to be approved in the United States in the summer of 2023, but this is something the market is not yet ready for.

Pearson also made some interesting considerations regarding value and affordability. Value is assessed by health assessment agencies and comprises health gains over a lifetime. Value is tempered by the uncertainty of effect durability and unknown safety risks, the judgment of whether cost offsets should be shared with society, and the balance between short-term affordability and long-term value. Affordability is the judgement based on implicit trade-offs within the health system, across all societal investments in well-being. Paths toward affordability include reducing the cost of production, competition, sharing of intellectual property, risk stratification, regulated pricing at launch, and payments over time linked to patient outcomes.[59]

Price is not the only barrier to equality. Equality as having access to the same opportunities is often used as a factor to determine fairness. This interpretation of equality is common in sports. "To effectively compare competitors' performance in a sporting contest, then, they need to be drawn from a single biological category, for instance considering sex, weight, age, etc."[60] Transhumanists have questioned "whether one might instead create two versions of the sport—one where enhancement is allowed and one where it is banned".[61] Natasha Vita-More, during the International Sport Accord Convention in St. Petersburg in 2013, introduced the

[58] Cf. «Third International Summit on Human Genome Editing», 2023, *The Royal Society*, in https://royalsociety.org/science-events-and-lectures/2023/03/2023-human-genome-editing-summit/ [7-3-2023].

[59] Cf. Pearson in *Ibid.*

[60] Cf. N. Bostrom - R. Roache, «Ethical Issues in Human Enhancement», in J. Ryberg - T. Petersen - C. Wolf (edd.), *New Waves in Applied Ethics, New Waves in Philosophy*, Palgrave Macmillan UK, Hampshire and New York 2007, 130.

[61] *Ibid.*, 132.

concept of a “Super Olympics”[62], as a way of including enhanced athletes in sports and respecting fairness. Can equality be ensured by adding a new category division?

Equality understood as having the same outcomes is another nuance. However, this interpretation of equality threatens diversity. Equal results are not always good and desired. Equality is desired in one sense, and difference is desired in another sense. Altogether, everyone should be respected as equal human beings while preserving the diversity of all those belonging to humanity. In short, the aim is to uphold equal moral worth and dignity, while saving diverse expressions of humanity.

9.4.1 Emphasising Intrinsic Equality, Extrinsic Diversity

Great attention is given to equality-solidarity in the selected WHO texts. Indeed, most of the values for decision-making have been related to equality and solidarity (*Inclusiveness, Fairness, Social justice, Non-discrimination, Equal moral worth, Solidarity* and *Global health justice*). This is certainly a good thing.

To ensure that equality and solidarity are put into practice, it is necessary to ensure that these values have a solid foundation. Otherwise, these values can be invoked to embellish actions, but then these actions might be empty and some people may be left behind.

If linked to equal moral worth, equality needs to be intrinsic. In other words, given to all human beings for the mere fact that they are humans. If the notion of equality is extrinsic, equality will never be real. “If human dignity is extrinsic, then not everyone is equal (…) There will always be individuals who manifest more or less of a particular trait.”[63] Austriaco rightly points out “There will always be individuals who are more intelligent, more conscious, more autonomous, more productive, or more aware of themselves, than others.”[64] In short, if moral worth depends on particular traits, there will be always someone more worthy than others. In such a scenario equality as equal moral worth would have no sense at all. There would be people who are well-treated, while others are mistreated. “Only an intrinsic account of human dignity based solely on the common humanity of all human beings can adequately explain the ideal of human equality that grounds a liberal society.”[65]

Moving forward, the objective is to refine the value of equality-solidarity emphasising “intrinsic equality, extrinsic diversity”. The equal distribution of healthcare and the equality of opportunities are goals that consequently emerge

---

[62] Cf. N. VITA- MORE, «About», *Natasha Vita-More PhD,* in https://natashavita-more.com/about/ [24-8-2021].

[63] N. P. G. AUSTRIACO, *Biomedicine and Beatitude: An Introduction to Catholic Bioethics*, The Catholic University of America Press, Washington, D.C. 2021(2), 379.

[64] *Ibid.*

[65] *Ibid.*

from intrinsic equality. It is a similar case for the basic need/human right to health care. This right does not exist because all people currently have access to healthcare, but rather it aims to ensure that healthcare for all because all humans are equal, have the same basic needs, and deserve health assistance.

To conclude this chapter, the commonalities that exist between H+ and WHO texts leave bioethicists with the task of refining the predominant values related to HGE issues. The first value, *individual rights and autonomy*, should refrain from simply meaning the fulfilment of personal wishes in health and reproduction to connect to other values such as humility, responsibility, and sociability. Also, to achieve liberty, careful consideration should be given to explicit and implicit coercions. The second value, *protecting people who cannot express their wishes*, should be broadened to include all people in need of protection and all the vulnerable. Lastly, the third value, *equality-solidarity*, should emphasise intrinsic equality to become consistent and fruitful.

Specifying the Ethics of Human Genome Editing

10

Abstract

This chapter aims to specify an ethical reasoning for human genome editing (HGE). First, it evidences how this reasoning cannot be as simple as confronting *somatic versus germline* HGE, and *therapy versus enhancement*. Concerning the last division, as shown by the principle of proportionality, intentions such as life preservation, health restoration, disease prevention, functional improvements, and personal preferences respond to different necessity levels. Concerning *somatic versus germline* HGE, recent documents refer to *somatic*, *germline* (not for reproduction) and *heritable* (for reproduction). To do a bioethical analysis, germline and heritable HGE must be subdivided into gene editing done in *gametes precursor cells* (female/male), *gametes* (female/male), and *embryos*. Precise identification of the type of HGE is essential for a complete bioethical analysis. Lastly, brief comments regarding the ethics of the various types of human genome editing are offered.

Keywords

Ethics • Bioethics • Human genome editing • Therapy • Enhancement

10.1 Important Differentiations and Subdivisions

10.1.1 From Therapy Versus Enhancement to Necessity Levels

At the end of this book, an attempt to specify the ethics of HGE is presented. We should depart from the usual division: *somatic versus germline* and *therapy versus enhancement*.

L. Santos, *Governance of Human Gene Editing and Transhumanism*,
Integrated Science 28, https://doi.org/10.1007/978-3-031-70863-3_10

Table 10.1 Somatic versus Germline and Therapy versus Enhancement

	Therapy	Enhancement
Somatic		
Germline		

Not long ago, most people would say that only somatic use of HGE for therapy purposes should be allowed. The germline use of HGE was seen as too risky, and often related to enhancement, therefore the general opinion was against it. Regarding therapeutic and enhancement uses, enhancement has long been considered unacceptable. It has been identified with twentieth-century eugenics with its devastating consequences. In addition, enhancement has been considered as seeking social advantage, or rebellion against what is naturally given. Therefore, somatic HGE for therapeutic uses was seen as the only acceptable scenario.

However, considering what was previously explained in this book, the Table 10.1 is unfit for ethical analysis. It is rather poor and insufficient. Regarding the therapy-enhancement division, as evidenced in Chap. 5 of this book, "therapy (allowed) versus enhancement (forbidden)" is a limited and falling paradigm, and it is inadequate to determine the ethics of an intervention. Chapter 5 has explained the problems caused by equivocal concepts such as health, the goals of medicine, enhancement, and normalcy. Subsequently, it has analysed some aspects that weaken "therapy versus enhancement", such as the goal of disease prevention (claimed by both therapy and enhancement), the improvement in quality of life for the elderly (above the average for this age), and some commonly used enhancements (not every action aiming at improvement is wrong). Lastly, the weakest point of "therapy versus enhancement" as a criterion for morality has been acknowledged: ethical reasoning based exclusively on intentions lacks consistency. It may cause significant harm in some cases while missing morally good or neutral improvements in other cases.

After the ethical assessment of the object, intentions, and circumstances of enhancement undertaken in Chap. 6, this book has proposed an improved framework for bioethical analysis, instead of *therapy versus enhancement.* This new framework, presented in Chap. 7, includes the principle of proportionality for the adequate choice among legitimate means for intervention.[1] The principle is not based on the therapeutic and enhancement intention to allow or prohibit the use of

[1] This framework is synthesised in the second part of the article: L. SANTOS, «Transhumanism and Global Governance of Human Genome Editing: Common Themes and Implications for Bioethics», *Medicina y Ética* 34/4 (2023), in https://doi.org/10.36105/mye.2023v34n4.05 [2-10-2023], 1089–1163.

any means. Also, it does not label a means, such as HGE, as inherently bad only because it could be misused. It recognises different necessity levels and means risk levels. The adequate choice among legitimate means for interventions implies proportionality in these terms:

- *Risky means are proportionally permitted to the level of necessity.* Riskier means are allowed at higher necessity levels because the benefit outweighs the risk in such cases. However, riskier means are not allowed at lower necessity levels, because the risk outweighs the benefit in such cases.
- *Low-risk means are allowed at any necessity level.*

The different levels of necessity for intervention relate to Maslow's pyramid of needs and respond to the basic human needs protected by human rights. The identified necessity levels are:

- *Life preservation*: actions aiming to save life (e.g., resuscitation).
- *Health Restoration*: actions aiming to achieve normal levels of health (e.g., hormonal treatment).
- *Disease Prevention*: actions aiming to avoid illness (e.g., vaccines).
- *Functional Improvements*: actions aiming to increase the performance of a useful trait (e.g., increasing concentration).
- *Preferences*: actions that produce no objective improvement, based only on desires and personal preferences (e.g., changing hair colour).

This requires a change in Table 10.1, replacing *Therapy* and *Enhancement* with the following categories:

Table 10.2 Necessity levels

Necessity levels (principle of proportionality)				
Life	Cure	Prevention	Functional	Preference

10.1.2 From Somatic Versus Germline to More Subdivisions

Back to Table 10.1, the column regarding the types of HGE (somatic and germlines) must also be modified. Germline HGE, which until a few years ago, was considered the only category different from Somatic HGE, is now subdivided into *germline* and *heritable* HGE. As previously explained in this book, the international scientific community and the governance documents now differentiate *somatic*, *germline* and *heritable* human genome editing.

The 2023 Statement from the Organising Committee of the Third Summit on HGE explains: "Somatic human genome editing refers to the editing of somatic cells, which are non-reproductive cells, and changes made in these cells affect only the person who receives the genome editing."[2] Regarding germline HGE, the Statement says: "germline human genome editing refers to the editing of human embryos or gametes in a *research setting*, with no plans for those embryos or gametes to be used for human reproduction."[3] In contrast, heritable HGE "refers to the editing of human embryos or gametes to be used *for human reproduction*."[4]

A nuance regarding germline and heritable HGE should be noted. While the Statement says that germline and heritable HGE is the editing of *embryos or gametes*, other important previous publications refer to germline and heritable HGE as the editing of *embryos, gametes or gametes precursor cells*. For instance, the WHO framework for governance states: "heritable human genome editing (that is, research that involves the transfer of edited *gamete precursor cells, gametes, or embryos* to a uterus with the aim of establishing a pregnancy)"[5] Similarly, the study on the global policy landscape for germline and heritable HGE says: "Like heritable genome editing, germline genome editing also involves making genetic modifications to *gamete precursor cells, eggs, sperm, or early-stage embryos* (…)".[6] Therefore, it is more correct to say that germline and heritable HGE is the

---

[2] The Organising Committee of the Third International Summit on Human Genome Editing, «Statement from the Organising Committee», 2023, in https://royalsociety.org/news/2023/03/statement-third-international-summit-human-genome-editing/ [7-3-2023].

[3] *Ibid.* The italic is mine.

[4] *Ibid.* The italic is mine.

[5] WHO Expert Advisory Committee on Developing Global Standards for Governance and Oversight of Human Genome Editing, *Human genome editing: a framework for governance*, WHO, Geneva 2021, in https://www.who.int/publications/i/item/9789240030060 [13-7-2021], 50. The italic is mine.

[6] F. Baylis et al., «Human Germline and Heritable Genome Editing: The Global Policy Landscape», *The CRISPR Journal* 3/5 (2020), in https://www.liebertpub.com/doi/10.1089/crispr.2020.0082 [14-7-2021], 367. The italic is mine.

gene editing of embryos, gametes or gametes precursor cells (germline HGE for research only, and heritable HGE for reproduction).

To correctly analyse an HGE intervention, it is not enough to use the general categories such as germline HGE or heritable HGE as they comprise both gametes, gametes precursor cells, and embryos. It is important to distinguish at least between human cells and human beings. Human cells, such as gametes and gamete precursor cells, are different from a new individual of the human species in the early stages of development. The human embryo differs from a cell of the male or female progenitor. The embryo depends on a number of conditions to fully develop into an *adult* human being, but not to become an *organism of the human species*.

There are five ways in which germline and heritable HGE can be performed: gene editing of *female and male precursor cells*, of *female and male gametes*, and of early-stage *embryos*. It should be noted that, as gametes and precursor cells are either male or female, they develop differently, and editing them implies different technical challenges. In the case of ex vivo editing, it should be noted that gamete precursor cells and gametes are harvested differently in men and women, having different implications for them.

Therefore, an improved table to support the ethical analysis should consider all these subdivisions regarding the type of HGE:

Table 10.3 Types of human genome editing

Somatic		
Germline	Gametes precursor cells	Female
		Male
	Gametes	Female
		Male
	Embryo	
Heritable	Gametes precursor cells	Female
		Male
	Gametes	Female
		Male
	Embryo	

10.1.3 A Complete Table for Ethical Analysis of HGE

As can be glimpsed, a table for bioethical analysis of HGE should be much more specific than the first table presented in this chapter (Table 10.1), which contrasts somatic and germline HGE and opposes therapy and enhancement to each other. To adequately choose among legitimate means (in this case, HGE), the many necessity levels (*life preservation, health restoration, prevention from diseases, functional improvements, personal preferences*) should be contemplated rather than making the simple division between therapy and enhancement. Also, instead of somatic and germline HGE, the different types of gene editing (*somatic, germline, heritable*) and the subdivisions of these categories (*female/male gametes precursor cells, female/male gametes, embryos*) should be considered. A complete table, (Table 10.4), would look like a combination of Tables 10.2 and 10.3.

Table 10.4 A complete table for ethical analysis

			Life	Cure	Prevention	Functional	Preference
Somatic							
Germline	Gametes precursor cells	Female					
		Male					
	Gametes	Female					
		Male					
	Embryo						
Heritable	Gametes precursor cells	Female					
		Male					
	Gametes	Female					
		Male					
	Embryo						

10.2 The Ethics of Somatic, Germline, and Heritable HGE

A complete bioethical analysis of a specific HGE should be done case by case to consider the proposed intervention's objective, intentions, and circumstances. However, at the general level, there are important bioethical considerations related to each type of HGE. These implications are displayed in Table 10.5 and briefly explained later on.

All bioethical analysis follows an ethical framework. The question is: which is the chosen ethics for Bioethics? Furthermore, what is the idea of humanity (anthropology) behind this ethical framework? Depending on the anthropological views, ethical assumptions, and bioethical principles, some bioethicists may reach different conclusions from the ones presented in the section. It is impossible to explain and justify here all the elements of the ethical framework used in this analysis. However, some fundamental ideas underlining this bioethical analysis should be made explicit:

- *Humans as embodied and social beings, and the resultant consequences*

Human nature or identity may be difficult to agree upon. But, while not forcing a complete definition, some facts can easily be observed. The human body forms a part of our human identity. Among other aspects, such as historicity and finitude, our physical body displays our special vulnerability and interdependency. This is especially evident in certain moments such as the beginning of life and infancy, moments of illness, and the end of life. The human body also speaks of both the need and the capacity for interrelationships, love, and care.[7]

- *Intrinsic dignity, equal moral worth, and respect for human rights*

All human beings are entitled to dignity. All humans have equal moral worth due to the mere fact of being a member of the human species. No discrimination has a place at this inherent and common level. Respect for human dignity is specified by the human rights. Human rights respond to human basic needs. All human beings have these rights from the moment they start existing, and until their life comes to an end. Considering the diversity of cultures and traditions, the basic rights stated in the *Universal Declaration of Human Rights*[8] are a helpful tool to respect every human being in the way they deserve. Also relevant to this work, the *Universal Declaration of Bioethics and Human Rights*[9] has linked Bioethics to human rights.

[7] Cf. O. C. Snead, *What It Means to Be Human*. Harvard University Press, Cambridge 2020.
[8] Cf. UN, «Universal Declaration of Human Rights», 1948, in https://www.ohchr.org/en/human-rights/universal-declaration/translations/english [13-7-2019].
[9] Cf. UNESCO, «Universal Declaration on Bioethics and Human Rights», in http://portal.unesco.org/en/ev.php-URL_ID=31058&URL_DO=DO_TOPIC&URL_SECTION=201.html [8-6-2019], 80.

- *Science and technology for humanity, and the role of Bioethics*

Humans are intelligent beings. Since pre-history, the human species has created tools and co-evolved with their technologies. The human capacity to understand reality and to transform it has shaped history. However, the natural capacity to create artificial means does not ensure that every discovery or invention will be helpful for humanity. History is also tainted by the misuse of intelligence and liberty. However, all approaches that blame technology and that try to erase human capacity for science should be avoided. It is necessary to find ways of developing and using technology according to our human identity. This means innovations that both respect and promote human rights, in order to fulfil human basic needs, which are the conditions for human flourishing. Bioethics is not just about pointing out the dangers of misusing technologies. It is also here to inspire, confirm, and encourage good uses of technology. Bioethics sets the direction for the high-tech marathon, for there is no point in running fast but in the wrong direction. Genuine progress is ethical and favours nature and humanity.

- *Personalist Bioethics and its principles*

Considering the various currents of thought in Bioethics, Personalism is an option that agrees with the previously cited views of humanity. The principles of Personalist Bioethics are: first, the *defence of physical life,* as one cannot exist, be free, and have human rights without life. Second: *liberty and responsibility,* which are connected in order to achieve individual and collective good. Third: *totality and therapy*, since the body is organically, unified, it is accepted to intervene in one part to save the whole organism. Lastly, *sociability and subsidiarity*: life is a social good, not just an individual one. Society must help without replacing individual initiatives, and it implies justice and solidarity.[10]

As there is no intention of analysing a specific case here, the next table (Table 10.5) skips the necessity levels presented in Table 10.4. It simply supposes high necessity levels and shows the types of HGE involved. In the column titled "Bioethics", Table 10.5 gives some examples of essentialist bioethical considerations, which are the ones related to the technique in itself, rather than the circumstantial considerations surrounding the intervention such as equality of access, informed consent, etc. The bioethical comments presented in Table 10.5 could be used in a more complete bioethical analysis.

Following Table 10.5, a brief explanation of the bioethical comments is now offered.

[10] Cf. E. SGRECCIA, *Manuale di bioetica I*, Vita e Pensiero, Milano 1994 (2), 171–180.

Table 10.5 Ethics of somatic, germline, and heritable HGE (considering the moratorium)

Human genome editing			Bioethics	
Somatic			OK	Proportional to the need
Germline	Primordial germ cells	Female	OK	Foetal development
		Male	OK	*Most promising ethical option
	Gametes	Female	OK	Increasing knowledge
		Male	OK	Increasing knowledge
	Embryo		X	Creating-Discarding embryos
Heritable	Primordial Germ Cells	Female	X	WHO Moratorium
		Male	X	
	Gametes	Female	X	
		Male	X	
	Embryo		X	

10.2.1 Somatic HGE (Non-reproductive Cells)

Somatic HGE (whether in uterus or after birth) presents no intrinsic ethical problem. Bioethical analysis can be done considering the object, end, and circumstances of a proposed genetic intervention, attentive to the proportionality between the necessity and risk level involved.

According to the Organising Committee of the Third International Summit on HGE, numerous clinical trials using somatic HGE are in progress, with preliminary but encouraging results for future therapies for both genetic and acquired diseases. The Statement from the Organising Committee also says further research to diversify and increase the efficiency, specificity, and safety of somatic HGE is essential.[11]

10.2.2 Germline HGE (Not for Reproduction)

i. *Germline HGE of Female and Male Gametes Precursor Cells*

The gametes precursor cells need to be extracted and the gene editing must be conducted outside the body, in the laboratory setting only. Otherwise, the edited germ cells could result in pregnancy, which would mean moving from germline HGE (for research only) to heritable HGE (for reproduction).

[11] Cf. The Organising Committee of the Third International Summit on Human Genome Editing, «Statement from the Organising Committee», 2023, in https://royalsociety.org/news/2023/03/statement-third-international-summit-human-genome-editing/ [7-3-2023].

- *Germline HGE of female gametes precursor cells.* Female gametes precursor cells are formed during the early stages of foetal development. The risks for the foetus and the pregnant woman should be considered. Rather than an intrinsic bioethical obstacle, there is a technical challenge.
- *Germline HGE of male gametes precursor cells.* Ethically speaking, this is the most promising option if heritable HGE is to be done in the future, whereby the man is a patient who would have his gametes precursor cells treated in order to produce healthy sperm. Germline HGE in male gametes precursor cells can increase knowledge.

ii *Germline HGE of Female and Male Gametes*

The gametes need to be harvested and the gene editing must be conducted outside the body, in a laboratory setting only. Otherwise, the edited gametes could result in pregnancy, which would mean moving from germline HGE (for research only) to heritable HGE (for reproduction).

- *Germline HGE of female gametes* (eggs) can advance research and contribute to knowledge, without affecting any child and with no risk of passing the genetic modification to the offspring.
- *Germline HGE of male gametes* (sperm) can advance research and contribute to knowledge, without affecting any child and with no risk of passing the genetic modification to the offspring.

iii *Germline HGE of Embryos*

Germline gene editing in embryos is done in the laboratory and for research only. After the completion of the research study, the edited embryos are discarded. These gene-edited embryos are never meant for reproduction (differently from heritable HGE).

- *Germline HGE of embryos,* while not prohibited in some countries[12] and even recommended in the Statement of the Third International Summit on Human Genome Editing,[13] is a highly controversial issue. Some defend reasons to edit human embryos and discard them later. Others, including myself, find no reason to justify intentionally eliminating these human beings. It is important to

[12] Cf. F. BAYLIS ET AL., «Human Germline and Heritable Genome Editing», The Global Policy Landscape», *The CRISPR Journal* 3/5 (2020), in https://www.liebertpub.com/doi/10.1089/crispr.2020.0082 [14-7-2021].

[13] Cf. THE ORGANISING COMMITTEE OF THE THIRD INTERNATIONAL SUMMIT ON HUMAN GENOME EDITING, «Statement from the Organising Committee», 2023, in https://royalsociety.org/news/2023/03/statement-third-international-summit-human-genome-editing/ [7-3-2023].

distinguish between human reproductive cells such as gametes and gametes precursors cells, and a new individual of the human species in the early stages of development.

10.2.3 Heritable HGE (For Reproduction)

Heritable HGE (HHGE) is currently not recommended in any circumstance. The case of CRISPR babies (2018) was followed by a moratorium (2019). WHO stated that "it would be irresponsible at this time for anyone to proceed with clinical applications of human germline genome editing."[14] Clinical applications of germline HGE is now called heritable HGE. WHO Director-general said he had "accepted the interim recommendations of WHO's Expert Advisory Committee that regulatory authorities in all countries should not allow any further work in this area until its implications have been properly considered."[15]

A moratorium is not a permanent ban, but a prohibition "at this time".[16] The reasons for the moratorium on HHGE are the current lack of safety, good governance, and social acceptability. For now, HHGE is too risky, governance of all types of gene editing is in its infancy, and there is insufficient social debate about genetic modifications that would be transmitted to future generations. It should be noted that even if reasonable safety is achieved to proceed with HHGE, long-term and intergenerational follow-up would be required.

However, none of the reasons for the moratorium are intrinsic and unchangeable. Although difficult to imagine today, the current stage of these three factors can change over time. In that hypothetical scenario, some ethical considerations regarding HHGE:

i. *Heritable HGE of Embryos*
   - *Heritable HGE of embryos.* Conducted in the laboratory and implying assisted reproductive technologies (ART), this type of HGE has the problem mentioned regarding germline editing in embryos, with the difference being that one or few more embryos would be implanted instead of discarded, resulting in birth.

ii. *Heritable HGE of Female and Male Gametes*

[14] WHO, «Statement on governance and oversight of human genome editing», 2019, in https://www.who.int/news/item/26-07-2019-statement-on-governance-and-oversight-of-human-genome-editing [13-7-2021].

[15] WHO, «Statement on governance and oversight of human genome editing», 2019, in https://www.who.int/news/item/26-07-2019-statement-on-governance-and-oversight-of-human-genome-editing [13-7-2021].

[16] WHO, «Statement on governance and oversight of human genome editing», 2019, in https://www.who.int/news/item/26-07-2019-statement-on-governance-and-oversight-of-human-genome-editing [13-7-2021].

- *Heritable HGE of female gametes.* Considering the gene editing of eggs outside the human body, pregnancy involves ART. This requires another proper ethical analysis, also considering the different types of ART and their nuances (gamete harvesting process, homologous/heterologous reproduction, anonymous or open gamete donation, surrogacy, fate of supernumerary embryos, etc.). Briefly, significant regard should be given to the dignity of all stakeholders, the health risks for the child and the pregnant women, the fate of embryos that are not implanted, the future family's well-being, and the best interest of the unborn child.
- *Heritable HGE of male gametes.* Considering the gene editing of sperm outside the human body, pregnancy then involves ART. The above considerations apply.

iii. *Heritable HGE of Female and Male Gametes Precursor Cells*

If gametes precursor cells were genetically edited, either in vivo (inside the body) or ex vivo (outside the body and then returned to it), the genetic alteration is expected to pass to the gametes. This would be a treatment of the patient's own cells, with potentially good consequences for the offspring, who could be naturally conceived with the healthy gamete.

If later ART is used aiming to check the embryo (created with gametes produced after genetically editing a gamete precursor cell), the previous considerations regarding the embryo and ART apply. However, in theory, heritable HGE of gametes precursor cells can be conducted and then pregnancy can occur naturally, no ART is required.

- *Heritable HGE of female gametes precursor cells.* As mentioned, female gametes precursor cells are formed during the early stages of foetal development. The risks for the foetus and the pregnant woman should be considered. Rather than an intrinsic bioethical obstacle, there is a technical challenge.
- *Heritable HGE of male gametes precursor cells.* This would be a treatment of the patient's own cells, with expected good consequences for his children, who could be naturally conceived. This would be the most ethical option if the conditions for lifting the moratorium are sufficiently met one day.

By doing this brief analysis on HHGE, this book is not recommending it, but doing a theoretical exercise regarding a hypothetical future scenario. This is characteristic of proactive Bioethics. If safety, good governance, and social acceptability are achieved, and the moratorium on HHGE is lifted, these comments will be key for ethical analysis.

Conclusion

Lessons Learned

This book identified and offered an initial work regarding some tasks for today's Bioethics, considering the common themes between the global governance of HGE and Transhumanism. This research (2018–2024) included many hours of lessons and readings, months applying the RTA method, numerous conversations with experts, and participation in academic conferences, workshops, public dialogues, and other gatherings such as the 2023 Third Summit on Human Genome Editing. However, the identified tasks for today's Bioethics are far from being concluded. Although completing these tasks is beyond my capabilities, over the years of this research, I have learned valuable lessons that I would like to share as a conclusion to this book.

Firstly, the *necessity of being part of the work on cutting-edge topics*. "Slopes are only slippery if they catch us unawares and we have strayed onto them inadequately equipped."[1] Technology is developing at an exponential rate. What seems impossible or too risky today, might be possible or feasible tomorrow. The way of handling these common issues is evolving and forming the framework for global governance of emerging technologies such as HGE. Society's mentalities and moral values are also changing. Bioethicists should be proactive, constructive, and proposal-oriented regarding issues that are still being shaped.

Secondly, we should *accept (and rejoice) that we are living "here and now"*. Instead of denying, evading, or only condemning the current challenges, it is better to depart from our reality and be constructive in creating solutions. We should embrace our here-and-now and strive towards the good of the individual and society. Nostalgia and fear might hinder bioethicists, as working with classical and clearer topics feels safer than working on new frontiers. It is necessary

[1] J. Harris in J. H. EVANS, *The Human Gene Editing Debate*, Oxford University Press, New York 2020, 18.

L. Santos, *Governance of Human Gene Editing and Transhumanism*, Integrated Science 28, https://doi.org/10.1007/978-3-031-70863-3

to accompany our society with its technologies, currents of thought, and social trends, instead of only judging when a praxis is already spread. It is better to be proactive than reactive in bioethical dilemmas.

Thirdly, there is a need for *constructive participation*. Bioethicists should offer a creative service translated into deeds. In our global society, this implies many dialogues, searching for convergences and proposals for cooperation. To constructively participate, we need to depart from the current situation and make proposals towards what we find to be ethical for the good of the individual and society.

Another lesson relates to the importance of *understanding the desires and fears of our time*. The problem is not the control over genes, but the goals we set and the means we choose to achieve them. "In this regard Transhumanism renders an invaluable service to humanity (…) transhumanism is forcing us to ask uncomfortable questions about our deepest desires, the means we use to achieve them, and the final outcome of all our technological transformations."[2]

Lastly, *sometimes it is necessary to find new ways to explain reality*. New praxis changes our vision and provokes questions. We should let ourselves be questioned. Transhumanism and HGE challenge our understanding of technology, biology, and humanity. This challenge is not necessarily bad. It can rather be a good opportunity. Transhumanism and HGE provide an invitation to find new ways to explain things that were taken for granted. Even true definitions can be improved. Indeed, new technologies, currents of thought, and social trends have the potential to shed light upon new aspects, urging us to increase our knowledge and find better ways to explain the same reality. So, thank you Transhumanism. Thank you, gene editing technologies. Thank you, governance of human genome editing.

[2] R. COLE- TURNER, *Transhumanism and Transcendence: Christian Hope in an Age of Technological Enhancement*, Georgetown University Press, Washington, D.C. 2011, 14.

Annexes

Interactive Presentations

I. The Applied RTA Method

The *presentation* graphically shows the six steps of the Reflexive Thematic Analysis (RTA) method used in the present research. This method was applied to selected texts on the governance of HGE and the selected texts on Transhumanism. Three common themes were found and developed. After that, some tasks for Bioethics were identified.

II. The Ethics of Human Genome Editing Course

This *presentation* is a didactic resource for a course related to the content of this book. This course introduces the current gene editing technologies and outlines the proposals for somatic, germline, and heritable HGE. It also examines the "therapy vs enhancement" paradigm. The course introduces the governance of HGE and

L. Santos, *Governance of Human Gene Editing and Transhumanism*, Integrated Science 28, https://doi.org/10.1007/978-3-031-70863-3

analyses different values and ethical arguments. Lastly, the students are encouraged to estimate the challenges and work on the tasks that HGE currently poses for Bioethics.

Bibliography

ACKERMANNARCHIVE, R., «Abortion pills via telemedicine: 10 Breakthrough Technologies 2023», 2023, *MIT Technology Review*, in https://www.technologyreview.com/2023/01/09/1064871/abortion-pills-telemedicine-10-breakthrough-technologies-2023/ [3-2-2023].

ALIGHIERI, D., *La Divina Commedia*, 1320, in https://divinacommedia.weebly.com/ [7-11-2022].

ANDORNO, R. - BAYLIS, F. - DARNOVSKY, M. - DICKENSON, D. - HAKER, H. - HASSON, K. - ET AL., «Geneva Statement on Heritable Human Genome Editing: The Need for Course Correction», *Trends in Biotechnology* 38/4 (2020), in https://www.sciencedirect.com/science/article/pii/S0167779919303178 [4-3-2021], 351–354.

ANZALONE, A. V. - RANDOLPH, P. B. - DAVIS, J. R. - SOUSA, A. A. - KOBLAN, L. W. - LEVY, J. M. - ET AL., «Search-and-replace genome editing without double-strand breaks or donor DNA», *Nature* 576/7785 (2019), in https://www.nature.com/articles/s41586-019-1711-4 [17-6-2021], 149–157.

AQUINAS, T., «Thomas de Aquino, Opera omnia», 2019, *Corpus Thomisticum*, in http://www.corpusthomisticum.org/iopera.html [12-12-2020].

ARAMESH, K., «Governance», in H. TEN HAVE (ed.), *Encyclopedia of Global Bioethics*, Springer International Publishing, Cham 2016, 1423–1432.

AUSTRIACO, N. P. G., *Biomedicine and Beatitude: An Introduction to Catholic Bioethics*, The Catholic University of America Press, Washington, D.C. 2021[2].

——, «Healthier than Healthy: The Moral Case for Therapeutic Enhancement», *The National Catholic Bioethics Quarterly* 17/1 (2017), in https://doi.org/10.5840/ncbq20171715 [5-7-2021], 43–49.

BALE, M., «Report: Workshop on Public Dialogue on Genomic Medicine», Council of Europe, Strasburg 25 April 2023, in https://rm.coe.int/cdbio-2023-1-rapporteur-report-on-public-dialogue-genomic-medicine-e/1680aae608 [1-5-2023], 20.

BALLOUZ, S. - DOBIN, A. - GILLIS, J. A., «Is it time to change the reference genome?», *Genome Biology* 20/1 (2019), in https://doi.org/10.1186/s13059-019-1774-4 [13-2-2020], 159.

BALTIMORE, D. - BERG, P. - BOTCHAN, M. - CARROLL, D. - CHARO, R. A. - CHURCH, G. - ET AL., «A prudent path forward for genomic engineering and germline gene modification», *Science* 348/6230 (2015), in https://www.ncbi.nlm.nih.gov/pmc/articles/PMC4394183/ [16-10-2020], 36–38.

BARABASI, A. L. - ALBERT, R., «Emergence of scaling in random networks», *Science* 286/5439 (1999), 509–512.

BAYLIS, F., *Altered Inheritance: CRISPR and the Ethics of Human Genome Editing*, Harvard University Press, Cambridge 2019.

——, «Human Genome Editing: Our Future Belongs to All of Us», *Issues in Science and Technology* XXXV/3 (2019), in https://issues.org/wp-content/uploads/2019/04/Baylis-Human-Genome-Editing-SP19-ISSUES-42-44.pdf [18-12-2019], 42–44.

BAYLIS, F. - DARNOVSKY, M. - HASSON, K. - KRAHN, T. M., «Human Germline and Heritable Genome Editing: The Global Policy Landscape», *The CRISPR Journal* 3/5 (2020), in https://www.liebertpub.com/doi/https://doi.org/10.1089/crispr.2020.0082 [14-7-2021], 365–377.

L. Santos, *Governance of Human Gene Editing and Transhumanism*, Integrated Science 28, https://doi.org/10.1007/978-3-031-70863-3

BEAUCHAMP, T. L. - CHILDRESS, J. F., *Principles of Biomedical Ethics*, Oxford University Press, New York 2001[5].

BENANTI, P., *Digital age. Teoria del cambio d'epoca. Persona, famiglia e società*, San Paolo Edizioni, Milano 2020.

—, «Il potenziamento cognitivo. Considerazione antropologiche ed etiche in prospettiva cristiana.», in G. QUARANTA (ed.), *Il doping della mente. Le sfide del potenziamento cognitivo farmacologico*, EMP, Padova 2014, 63–96.

—, *The Cyborg: corpo e corporeità nell'epoca del post-umano*, Studi e ricerche. Sez. teologica, Cittadella, Città di Castello 2012.

BERG, P. - BALTIMORE, D. - BRENNER, S. - ROBLIN, R. O. - SINGER, M. F., «Summary statement of the Asilomar conference on recombinant DNA molecules.», *Proceedings of the National Academy of Sciences of the United States of America* 72/6 (1975), in https://www.ncbi.nlm.nih.gov/pmc/articles/PMC432675/ [10-6-2022], 1981–1984.

BERNABE, R. D. L. C. - VAN THIEL, G. J. M. W. - RAAIJMAKERS, J. A. M. - VAN DELDEN, J. J. M., «The risk-benefit task of research ethics committees: An evaluation of current approaches and the need to incorporate decision studies methods», *BMC Medical Ethics* 13/1 (2012), in https://doi.org/10.1186/1472-6939-13-6 [21-1-2023], 6.

BOSTROM, N., «A History of Transhumanist Thought», in M. RECTENWALD - L. CARL (edd.), *Academic Writing Across the Disciplines*, Pearson Longman, New York 2011.

—, «Dignity and Enhancement», in *Human Dignity and Bioethics: Essays Commissioned by the President's Council on Bioethics*, President's Council on Bioethics, Washington, D.C. 2008, 173–206.

—, «Human Genetic Enhancements: A Transhumanist Perspective», *The Journal of Value Inquiry* 37/4 (2003), in https://doi.org/10.1023/B:INQU.0000019037.67783.d5 [12-7-2021], 493–506.

—, «In Defense of Posthuman Dignity», *Bioethics* 19/3 (2005), in https://onlinelibrary.wiley.com/doi/abs/https://doi.org/10.1111/j.1467-8519.2005.00437.x [10-4-2021], 202–214.

—, «Letter from Utopia», *Studies in Ethics, Law, and Technology* 2/1 (2008), in https://www.nickbostrom.com/utopia.html [12-7-2021], 1–7.

—, «Nick Bostrom's Home Page», in https://www.nickbostrom.com/ [12-7-2020].

—, «The Future of Humanity», *Geopolitics, History, and International Relations* 1/2 (2009), in https://www.nickbostrom.com/papers/future.html [12-7-2021], 41–78.

—, «Transhumanism: The World's Most Dangerous Idea?», 2004, in https://www.nickbostrom.com/papers/dangerous.html [7-10-2020].

—, «Transhumanist Ethics», in https://nickbostrom.com/ethics/transhumanist.pdf [10-6-2020].

—, «Transhumanist Values», *Journal of Philosophical Research* 30/9999 (2005), 3–14.

—, «What is transhumanism?», 2001, in https://www.nickbostrom.com/old/transhumanism.html [15-9-2020].

BOSTROM, N. - ROACHE, R., «Ethical Issues in Human Enhancement», in J. RYBERG - T. PETERSEN - C. WOLF (edd.), *New Waves in Applied Ethics*, New Waves in Philosophy, Palgrave Macmillan UK, Hampshire and New York 2007, in https://www.palgrave.com/gp/book/9780230537835 [12-7-2021], 120–152.

BOSTROM, N. - SANDBERG, A., «Cognitive Enhancement: Methods, Ethics, Regulatory Challenges», *Science and Engineering Ethics* 15/3 (2009), in https://doi.org/10.1007/s11948-009-9142-5 [19-11-2022], 311–341.

—, «The Wisdom of Nature: An Evolutionary Heuristic for Human Enhancement», in J. SAVULESCU - N. BOSTROM (edd.), *Human Enhancement*, OUP Oxford, Oxford 2011, 375–416.

BOSTROM, N. - SAVULESCU, J., «Human Enhancement Ethics: The State of the Debate», (2009), 1–22.

BOYLE, JOSEPH M., «Toward Understanding the Principle of Double Effect», *Ethics* 90/4 (1980), in https://www.journals.uchicago.edu/doi/abs/https://doi.org/10.1086/292183 [16-3-2023], 527–538.

BRAUN, V. - CLARKE, V., «Conceptual and design thinking for thematic analysis», *Qualitative Psychology* 9 (2022), 3–26.

——, «One size fits all? What counts as quality practice in (reflexive) thematic analysis?», *Qualitative Research in Psychology* 18/3 (2021), in https://doi.org/10.1080/14780887.2020.1769238 [11-11-2022], 328–352.

——, «Reflecting on reflexive thematic analysis», *Qualitative Research in Sport, Exercise and Health* 11/4 (2019), in https://doi.org/10.1080/2159676X.2019.1628806 [11-11-2022], 589–597.

——, *Successful Qualitative Research: A Practical Guide for Beginners*, SAGE, Los Angeles 2013.

——, «Thematic analysis», in *APA handbook of research methods in psychology, Vol 2: Research designs: Quantitative, qualitative, neuropsychological, and biological*, APA handbooks in psychology, American Psychological Association, Washington, D.C. 2012, 57–71.

——, *Thematic Analysis: A Practical Guide*, SAGE Publications Ltd, London and Thousand Oaks 2021.

——, «Using thematic analysis in psychology», *Qualitative Research in Psychology* 3/2 (2006), in https://www.tandfonline.com/doi/abs/https://doi.org/10.1191/1478088706qp063oa [31-8-2022], 77–101.

BREY, P. - NAGEL, S., «Bioengeenering», in H. TEN HAVE (ed.), *Encyclopedia of Global Bioethics*, Springer International Publishing, Cham 2016, 280–291.

BRICKLEY, L., «Bodies without Borders: The Sinews and Circuitry of "folklore+"», *Western Folklore* 78/1 (2019), in https://www.jstor.org/stable/26864140 [10-7-2021], 5–38.

BYK, C., «Transhumanism: from Julian Huxley to UNESCO: What Objective for International Action?», *Jahr - European Journal of Bioethics* 12/1 (2021), in https://www.jahr-bioethics-journal.com/index.php/JAHR/article/view/565 [15-11-2022], 141–162.

BYRNE, D., «A worked example of Braun and Clarke's approach to reflexive thematic analysis», *Quality & Quantity* 56/3 (2022), in https://doi.org/10.1007/s11135-021-01182-y [18-10-2022], 1391–1412.

CAMPA, R., *Mutare o perire. La sfida del transumanesimo*, Sestante Edizioni, Bergamo 2010.

CENTER FOR GENETICS AND SOCIETY, «Welcome to CGS», in https://www.geneticsandsociety.org/ [2-7-2021].

CLARK, R. W., *The Huxleys*, Heinemann, London 1968.

COENEN, C., «Transcending Natural Limitations: The Military-Industrial Complex and the Transhumanist Temptation», in W. HOFKIRCHNER - H.- J. KREOWSKI (edd.), *Transhumanism: The Proper Guide to a Posthuman Condition or a Dangerous Idea?*, Cognitive Technologies, Springer International Publishing, Cham 2021, 97–110.

COLE- TURNER, R., *Transhumanism and Transcendence: Christian Hope in an Age of Technological Enhancement*, Georgetown University Press, Washington, D.C. 2011.

COMMITTEE ON BIOETHICS OF THE COUNCIL OF EUROPE, «Genome editing technologies: some clarifications but no revision of the Oviedo Convention», 2021, *Human Rights and Biomedicine*, in https://www.coe.int/en/web/bioethics/news/-/asset_publisher/EV74osp47zWZ/content/genome-editing-technologies-some-clarifications-but-no-revision-of-the-provisions-of-the-oviedo-convention [12-7-2022].

CORTINA, A., *Humanismo avanzado para una sociedad biotecnológica*, EIUNSA, Madrid 2017.

COUNCIL OF EUROPE, «Convention for the Protection of Human Rights and Dignity of the Human Being with regard to the Application of Biology and Medicine: Convention on Human Rights and Biomedicine», European Treaty Series 1997, in https://rm.coe.int/CoERMPublicCommonSearchServices/DisplayDCTMContent?documentId=090000168007cf98 [15-7-2021].

——, «Workshop on public dialogue on genomic medicine», 2022, *Human Rights and Biomedicine*, in https://www.coe.int/en/web/bioethics/workshop-public-dialogue-genomic-medicine [10-11-2022].

CRESPO- RODRÍGUEZ, M. A., *Zoltan Istvan y el Partido Transhumanista: Política y transhumanismo en el siglo XXI*, University of Puerto Rico, Rio Piedras 2018, in https://www.proquest.com/docview/2023807738/abstract/B0C097A17A8F452BPQ/1 [12-5-2021].

DEBAETS, A. M., «Rapture of the Geeks: Singularitarianism, Feminism, and the Yearning for Transcendence», in *Religion and Transhumanism: The Unknown Future of Human Enhancement*, ABC-CLIO, Santa Barbara 2014, 181–197.

DELFANTI, A., *Biohackers. The Politics of Open Science*, Pluto Press, London 2013.

DELHOVE, J. - OSENK, I. - PRICHARD, I. - DONNELLEY, M., «Public Acceptability of Gene Therapy and Gene Editing for Human Use: A Systematic Review», *Human Gene Therapy* 31/1–2 (2020), in https://www.liebertpub.com/doi/https://doi.org/10.1089/hum.2019.197 [5-7-2021], 20–46.

DESMOND, A. J., «Thomas Henry Huxley», in *Encyclopedia Britannica*, Philosophy & Religion, Encyclopedia Britannica, Chicago 2021, in https://www.britannica.com/biography/Thomas-Henry-Huxley [20-4-2021].

DIVALL, C. M., «From a Victorian to a Modern: Julian Huxley and the English Intellectual Climate», in C. K. WATERS - A. VAN HELDEN (edd.), *Julian Huxley: Biologist and Statesman of Science*, Rice University Press, Houston 1992, 31–44.

DOUDNA, J. A. - STERNBERG, S. H., *A Crack in Creation: Gene Editing and the Unthinkable Power to Control Evolution*, Houghton Mifflin Harcourt, Boston 2017.

EDWARDS, G., «Enhancement», in H. TEN HAVE (ed.), *Encyclopedia of Global Bioethics*, Springer International Publishing, Cham 2016, 1121–1130.

ENGELHARDT, H. T., «Morality, Universality, and Particularity: Rethinking the Role of Community in the Foundations of Bioethics», in J. T. L. PO- WAH (ed.), *Cross-Cultural Perspectives on the (Im)Possibility of Global Bioethics*, Philosophy of Medicine, Springer, Dordrecht 2002, in https://doi.org/10.1007/978-94-017-1195-1_2 [29-12-2022], 19–38.

ESFANDIARY, F. M., *Optimism one: The emerging radicalism*, Norton, New York 1970.

EUROPEAN GROUP ON ETHICS IN SCIENCE AND NEW TECHNOLOGIES, *Opinion on the Ethics of Genome Editing*, Publications Office of the European Union, Brussels 2021, in http://op.europa.eu/en/web/eu-law-and-publications/publication-detail/-/publication/6d9879f7-8c55-11eb-b85c-01aa75ed71a1 [16-11-2021].

—, *Values for the Future: The Role of Ethics in European and Global Governance*, Publications Office of the European Union, Brussels 2021, in https://ec.europa.eu/info/files/values-future-role-ethics-european-and-global-governance_en [16-11-2021].

EUROPEAN PARLIAMENT'S SCIENCE AND TECHNOLOGY OPTIONS ASSESSMENT, «Human Enhancement. Panel for the Future of Science and Technology», 2009, in https://www.europarl.europa.eu/stoa/en/document/IPOL-JOIN_ET(2009)417483 [17-6-2021].

EVANS, J. H., *Playing God? Human Genetic Engineering and the Rationalization of Public Bioethical Debate*, University of Chicago Press, Chicago 2002.

—, *The History and Future of Bioethics: A Sociological View*, Oxford University Press, New York 2014.

—, *The Human Gene Editing Debate*, Oxford University Press, New York 2020.

FERRETTI, F., «Mapping do-it-yourself science», *Life Sciences, Society and Policy* 15/1 (2019), in https://doi.org/10.1186/s40504-018-0090-1 [14-3-2023], 1.

FINNIS, J., *Natural Law and Natural Rights*, OUP Oxford, Oxford 2011.

FINS, J. J. - MILLER, F. G., «Proportionality, Pandemics, and Medical Ethics», *The American Journal of Medicine* 133/11 (2020), in https://www.ncbi.nlm.nih.gov/pmc/articles/PMC7347467/ [26-4-2022], 1243–1244.

FLEISCHHAUER, K. - HERMEREN, G., *Goals of Medicine in the Course of History and Today: A Study in the History and Philosophy of Medicine*, Stockholm 2006.

FORGE, J., «Proportionality, Just War Theory and Weapons Innovation», *Science and Engineering Ethics* 15/1 (2009), 25–38.

FUKUYAMA, F., *Our Posthuman Future: Consequences of the Biotechnology Revolution*, Profile Books, New York 2003.

GARLAND- THOMSON, S. S., ROSEMARIE, «The Dark Side of CRISPR», 2021, *Scientific American*, in https://www.scientificamerican.com/article/the-dark-side-of-crispr/ [3-3-2021].

GATES, A. J. - GYSI, D. M. - KELLIS, M. - BARABÁSI, A.- L., «A wealth of discovery built on the Human Genome Project — by the numbers», *Nature* 590/7845 (2021), in https://www.nature.com/articles/d41586-021-00314-6 [21-5-2021], 212–215.

GEHLEN, A., *Studien zur Anthropologie und Soziologie*, Soziologische Texte 17, Luchterhand Verlag, Neuwied 1963.

GIANNNKOULOPOULUS, A. - LIMNIATI, L. - HONORATO, D., *Portrayal of Biohackers in news and technological websites*, in (2017), 292–298.

GLOBAL OBSERVATORY FOR GENOME EDITING, «Global Observatory», 2020, in https://global-observatory.org/about/ [15-7-2021].

GRAY, C. H. - FIGUEROA- SARRIERA, H. - MENTOR, S., *Modified: Living as a Cyborg*, Routledge, New York 2020, in https://doi.org/10.4324/9781351107839 [12-12-2020].

GREELY, H. T., «How Should Science Respond to CRISPR'd Babies?», *Issues in Science and Technology* XXXV/3 (2019), in https://issues.org/how-should-science-respond-to-crisprd-babies/ [18-12-2019], 32–37.

GREEN, B. P., «Transhumanism and Catholic Natural Law: Changing Human Nature and Changing Moral Norms», in *Religion and Transhumanism: The Unknown Future of Human Enhancement*, ABC-CLIO, Santa Barbara 2014, 201–215.

GRISEZ, G. - SHAW, R., *Beyond the New Morality: The Responsibilities of Freedom*, University of Notre Dame Press, Notre Dame 1988^3.

GRISEZ, G. G., *Christian Moral Principles: Way of the Lord Jesus*, I, Franciscan Pr, Chicago 1983.

GRUBER, K., «Biohackers», *EMBO reports* 20/6 (2019), in https://www.embopress.org/doi/full/https://doi.org/10.15252/embr.201948397 [27-11-2020], e48397.

GUERRINI, C. J. - TREJO, M. - CANFIELD, I. - MCGUIRE, A. L., «Core values of genomic citizen science: results from a qualitative interview study», *BioSocieties* 17/2 (2022), 203–228.

GYNGELL, C. - BOWMAN- SMART, H. - SAVULESCU, J., «Moral reasons to edit the human genome: picking up from the Nuffield report», *Journal of Medical Ethics* 45/8 (2019), in https://jme.bmj.com/lookup/doi/https://doi.org/10.1136/medethics-2018-105084 [19-12-2020], 514–523.

HABERMAS, J., *The Future of Human Nature*, Polity Press, Cambridge 2003.

HALL, M., *The Bioethics of Enhancement: Transhumanism, Disability, and Biopolitics*, Lexington Books, Lanham and London 2016.

HALPERN, J. - O'HARA, S. E. - DOXZEN, K. W. - WITKOWSKY, L. B. - OWEN, A. L., «Societal and Ethical Impacts of Germline Genome Editing: How Can We Secure Human Rights?», *The CRISPR Journal* 2/5 (2019), in https://www.liebertpub.com/doi/full/https://doi.org/10.1089/crispr.2019.0042 [4-3-2021], 293–298.

HAMZELOUARCHIVE, J., «CRISPR for high cholesterol: 10 Breakthrough Technologies 2023», 2023, *MIT Technology Review*, in https://www.technologyreview.com/2023/01/09/1064857/crispr-high-cholesterol-10-breakthrough-technologies-2023/ [3-2-2023].

—, «Next up for CRISPR: Gene editing for the masses?», 2023, *MIT Technology Review*, in https://www.technologyreview.com/2023/01/19/1067074/next-for-crispr/ [3-2-2023].

HARRIS, J., «Enhancements are a Moral Obligation», in J. SAVULESCU - N. BOSTROM (edd.), *Human Enhancement*, OUP Oxford, Oxford 2011, 131–154.

TEN HAVE, H., *Global Bioethics: An Introduction*, Routledge, New York 2016.

HERMERÉN, G., «The principle of proportionality revisited: interpretations and applications», *Medicine, Health Care and Philosophy* 15/4 (2012), 373–382.

HIPPOCRATES, *Hippocratic Writings*, edd. J. CHADWICK - G. E. R. LLOYD - W. N. MANN, Penguin classics, Penguin Books, Harmondswerth 1983.

HOPKINS, P. D., «A Salvation Paradox for Transhumanism: Saving You versus Saving You», in *Religion and Transhumanism: The Unknown Future of Human Enhancement: The Unknown Future of Human Enhancement*, ABC-CLIO, Santa Barbara 2014, 71–81.

HUGHES, J. J., *Citizen Cyborg: Why Democratic Societies Must Respond To The Redesigned Human Of The Future*, Basic Books, Cambridge 2004.

—, «Democratic Transhumanism 2.0», (2002), in http://www.changesurfer.com/Acad/DemocraticTranshumanism.htm [10-7-2021].

—, «The Politics of Transhumanism and the Techno-Millennial Imagination, 1626–2030», *Zygon: Journal of Religion & Science* 47/4 (2012), in https://doi.org/10.1111/j.1467-9744.2012.01289.x, 757–776.

HUMANITY+, «Leadership», *Humanity+*, in https://www.humanityplus.org/leadership [14-10-2022].

—, «Mission», *Humanity+*, in https://www.humanityplus.org/about [11-6-2022].

——, «The Transhumanist Declaration», 2009, *Humanity+*, in https://www.humanityplus.org/the-transhumanist-declaration [11-6-2022].

——, «Transhumanism», *Humanity+*, in https://www.humanityplus.org/transhumanism [11-6-2022].

——, «Transhumanist FAQ 3.0», 2001, *Humanity+*, in https://www.humanityplus.org/transhumanist-faq [2-12-2020].

HURLBUT, J. B., «Imperatives of Governance: Human Genome Editing and the Problem of Progress», *Perspectives in Biology and Medicine* 63/1 (2020), in https://muse.jhu.edu/article/748059 [24-5-2022], 177–194.

HURLBUT, J. B. - SAHA, K. - JASANOFF, S., «CRISPR Democracy: Gene Editing and the Need for Inclusive Deliberation», *Issues in Science and Technology* XXXII/1 (2015), in https://issues.org/crispr-democracy-gene-editing-inclusive-deliberation/ [17-7-2021].

HUXLEY, A., *Brave New World*, Chatto & Windus, London 1932, in https://gutenberg.ca/ebooks/huxleya-bravenewworld/huxleya-bravenewworld-00-h.html [25-6-2020].

HUXLEY, J., *Religion Without Revelation*, Harper Brothers Publishers, New York and London 1927, in http://archive.org/details/in.ernet.dli.2015.90330 [15-5-2020].

——, «Transhumanism», *Journal of Humanistic Psychology - J HUM PSYCHOL* 8 (1968), 73–76.

HUXLEY, J. S. - PIGLIUCCI, M. - MULLER, G. B., *Evolution, The Definitive Edition: The Modern Synthesis*, The MIT Press, Cambridge 2009.

INTERNATIONAL BIOETHICS COMMITTEE (IBC), «Report of the IBC on updating its reflection on the Human Genome and Human Rights», International Bioethics Committee, Paris 2 October 2015, in https://unesdoc.unesco.org/ark:/48223/pf0000233258 [14-2-2020], 30.

INTERNATIONAL SOCIETY FOR STEM CELL RESEARCH, «ISSCR Guidelines for Stem Cell Research and Clinical Translation», May 2021, in https://www.isscr.org/policy/guidelines-for-stem-cell-research-and-clinical-translation [22-6-2021], 73.

ISTVAN, Z., «Expectations and Apprehensions on Transhumanism», *FIfF-Kommunikation 2/2016* 2/16 (2016), 19–23.

——, *Philosophy and the Future: A Transhumanist Examination of Where We're Going*, Rudi Ventures LLC 2020.

——, *The Anti-Deathist: Writings of a Radical Longevity Activist*, Futurity Imagine Media LLC 2020.

——, *Upgrading America: The Political Writings of Zoltan Istvan*, Rudi Ventures LLC 2019.

JASANOFF, S., *Can Science Make Sense of Life?*, New Human Frontiers, Polity Press, Cambridge 2019.

——, *The Ethics of Invention: Technology and the Human Future*, W. W. Norton & Co Inc, New York 2016.

JASANOFF, S. - HURLBUT, J. B., «A global observatory for gene editing», *Nature* 555/7697 (2018), in https://www.nature.com/articles/d41586-018-03270-w [24-5-2022], 435–437.

JULIAN HUXLEY, *New Bottles For New Wine*, Chatto and Windus LTD., London 1957, in http://archive.org/details/NewBottlesForNewWine [15-5-2020].

KARLOVITZ, T. J., «The Democratization of Technology – and Its Limitation», in T. DIRSEHAN (ed.), *Managing Customer Experiences in an Omnichannel World: Melody of Online and Offline Environments in the Customer Journey*, Emerald Publishing Limited 2020, in https://doi.org/10.1108/978-1-80043-388-520201004 [27-11-2020], 13–25.

KASS, L., «Preventing a Brave New World», in *Ethics and Emerging Technologies*, Palgrave Religion & Philosophy Collection, Palgrave Macmillan, London 2014, in https://web.stanford.edu/~mvr2j/sfsu09/extra/Kass3.pdf [7-3-2020], 76–89.

KELLMEYER, P., *Proportionality as a Guiding Principle for Regulating Invasive Neurotechnological Medical Devices*, in (2017), in https://www.ieeg.uni-freiburg.de/data/files/pdf/kellmeyer-INS2/view?searchterm=proporti [26-4-2022], poster 58.

KIGER, M. E. - VARPIO, L., «Thematic analysis of qualitative data: AMEE Guide No. 131», *Medical Teacher* 42/8 (2020), in https://doi.org/10.1080/0142159X.2020.1755030 [31-8-2022], 846–854.

KLEIDERMAN, E. - RAVITSKY, V. - KNOPPERS, B. M., «The 'serious' factor in germline modification», *Journal of Medical Ethics* 45/8 (2019), in https://www.ncbi.nlm.nih.gov/pmc/articles/PMC6820154/ [14-7-2021], 508–513.

KORFF, D., «The Right to Life: A Guide to the Implementation of Article 2 of the European Convention on Human Rights», (2008), in https://papers.ssrn.com/abstract=1288555 [8-11-2022].

KURZWEIL, R., *How to Create a Mind: The Secret of Human Thought Revealed*, Penguin Books, New York 2013.

—, *The Age of Spiritual Machines: When Computers Exceed Human Intelligence*, Penguin Books, New York 2000.

—, *The Singularity Is Near: When Humans Transcend Biology*, Penguin Books, New York 2005.

LANDER, E. S. - BAYLIS, F. - ZHANG, F. - CHARPENTIER, E. - BERG, P. - BOURGAIN, C. - ET AL., «Adopt a moratorium on heritable genome editing», *Nature* 567/7747 (2019), in https://www.nature.com/articles/d41586-019-00726-5 [17-6-2021], 165–168.

LEDFORD, H., «Biohackers gear up for genome editing», *Nature News* 524/7566 (2015), in http://www.nature.com/news/biohackers-gear-up-for-genome-editing-1.18236 [27-11-2020], 398.

LEE, N., «Brave New World of Transhumanism», in N. LEE (ed.), *The Transhumanism Handbook*, Springer International Publishing, Cham 2019, 3–48.

LEE, R., «Biohacktivism, Cyborgasm, and Transhumanism», in N. LEE (ed.), *The Transhumanism Handbook*, Springer International Publishing, Cham 2019, 499–501.

LI, H. - YANG, Y. - HONG, W. - HUANG, M. - WU, M. - ZHAO, X., «Applications of genome editing technology in the targeted therapy of human diseases: mechanisms, advances and prospects», *Signal Transduction and Targeted Therapy* 5/1 (2020), in https://www.nature.com/articles/s41392-019-0089-y [9-5-2022], 1–23.

LIANG, P. - XU, Y. - ZHANG, X. - DING, C. - HUANG, R. - ZHANG, Z. - ET AL., «CRISPR/Cas9-mediated gene editing in human tripronuclear zygotes», *Protein & Cell* 6/5 (2015), in https://doi.org/10.1007/s13238-015-0153-5 [11-11-2022], 363–372.

LILLEY, S., *Transhumanism and Society: The Social Debate over Human Enhancement*, SpringerBriefs in Philosophy, Springer, Heidelberg 2013, in https://www.springer.com/it/book/9789400749801 [20-5-2021].

MACINTYRE, A., *Whose Justice? Which Rationality?*, University of Notre Dame Press, Notre Dame 1988.

MANGAN, J., «An Historical Analysis of the Principle of Double Effect», *Theological Studies* 10 (1949), in https://doi.org/10.1177/004056394901000102 [16-3-2023], 41–61.

MANZOCCO, R., *Transhumanism - Engineering the Human Condition: History, Philosophy and Current Status*, Springer, New York 2019.

MARCHANT, G. E., «Global Governance of Human Genome Editing: What Are the Rules?», *Annual Review of Genomics and Human Genetics* 22/1 (2021), in https://www.annualreviews.org/doi/https://doi.org/10.1146/annurev-genom-111320-091930 [16-9-2021], 385–405.

MARCHIONE, M., «Chinese researcher claims first gene-edited babies», *AP NEWS* (2018), in https://apnews.com/article/4997bb7aa36c45449b488e19ac83e86d [19-12-2020].

MARTIN, A. R. - KANAI, M. - KAMATANI, Y. - OKADA, Y. - NEALE, B. M. - DALY, M. J., «Clinical use of current polygenic risk scores may exacerbate health disparities», *Nature Genetics* 51/4 (2019), 584–591.

MARTINHO DA SILVA, P., «Bioethics: And Politics», in H. TEN HAVE (ed.), *Encyclopedia of Global Bioethics*, Springer International Publishing, Cham 2016, 291–296.

MASLOW, A. H., «A theory of human motivation», *Psychological Review* 50/4 (1943), 370–396.

MASLOW, A. H., *Motivation and personality*, Harper's psychological series, Harper & Row, Publishers, New York 1954.

—, *Motivation and personality*, Harper & Row, New York 1970^2.

—, *Religions, values, and peak-experiences*, Penguin Arkana, New York 1994.

MASLOW, A. H. - COX, R., *Motivation and personality*, edd. R. FRAGER - J. FADIMAN - C. MCREYNOLDS, Pearson Education, Inc., Uttar Pradesh, India 1987^3.

MAYOR ZARAGOZA, F., «Dignity as the foundation for all Human Rights», in G. SOLINÍS (ed.), *Global bioethics: what for? Twentieth anniversary of UNESCO's Bioethics Programme*, UNESCO, Paris 2015, 47–51.

MENDZ, G. - COOK, M., «Transhumanist Genetic Enhancement: Creation of a 'New Man' Through Technological Innovation», *The New Bioethics: a multidisciplinary journal of biotechnology and the body* 27/2 (2021), 105–126.

MEYER, M. N. - TAN, T. - BENJAMIN, D. J. - LAIBSON, D. - TURLEY, P., «Public views on polygenic screening of embryos», *Science* 379/6632 (2023), in https://www.science.org/doi/https://doi.org/10.1126/science.ade1083 [20-2-2023], 541–543.

MIKKELSEN, R. B. - FREDERIKSEN, H. R. S. - GJERRIS, M. - HOLST, B. - HYTTEL, P. - LUO, Y. - ET AL., «Genetic Protection Modifications: Moving Beyond the Binary Distinction Between Therapy and Enhancement for Human Genome Editing», *The CRISPR Journal* 2/6 (2019), in https://www.liebertpub.com/doi/abs/https://doi.org/10.1089/crispr.2019.0024 [4-3-2021], 362–369.

MILLS, P., «Human genome editing and moral leadership: findings of the WHO Expert Advisory Committee», 2021, *The Nuffield Council on Bioethics*, in https://www.nuffieldbioethics.org/blog/human-genome-editing-and-moral-leadership-findings-of-the-who-expert-advisory-committee [14-7-2021].

—, «Three Venues for Discussing Human Gene Editing», *Issues in Science and Technology* XXXV/3 (2019), in https://issues.org/three-venues-for-discussing-human-gene-editing/ [18-12-2019], 38–41.

MORE, M., «Max More», in https://www.maxmore.com/ [14-11-2022].

—, «Philosophy of Transhumanism», *Humanity+*, in https://www.humanityplus.org/philsophy-of-transhumanism [19-11-2022].

—, *The Proactionary Principle*, in (2004), in http://www.extropy.org/proactionaryprinciple.htm [8-6-2021].

MOSSBRIDGE, J. A., «Leave Nobody Behind: The Future of Societies», in A. STOEL (ed.), *The Transhumanism Handbook*, Springer International Publishing, Cham 2019, 613–617.

MUSUNURU, K. - CHADWICK, A. C. - MIZOGUCHI, T. - GARCIA, S. P. - DENIZIO, J. E. - REISS, C. W. - ET AL., «In vivo CRISPR base editing of PCSK9 durably lowers cholesterol in primates», *Nature* 593/7859 (2021), in https://www.nature.com/articles/s41586-021-03534-y [14-3-2023], 429–434.

NATIONAL ACADEMIES OF SCIENCES, ENGINEERING, AND MEDICINE, «Human Genome Editing: Science, Ethics, and Governance», The National Academies Press, Washington, D.C. 14 February 2017, in https://doi.org/10.17226/24623 [5-7-2021], 329.

—, «International Summit on Human Gene Editing: A Global Discussion», *The National Academies Press* (2015), in https://doi.org/10.17226/21913 [15-7-2021], 8.

—, «Second International Summit on Human Genome Editing: Continuing the Global Discussion: Proceedings of a Workshop–in Brief», *The National Academies Press* (2019), in https://doi.org/10.17226/25343 [15-7-2021], 10.

NATIONAL ACADEMY OF SCIENCES - NATIONAL ACADEMIES OF MEDICINE - THE ROYAL SOCIETY, «Heritable Human Genome Editing», The National Academies Press, Washington, D.C. 3 September 2020, in https://doi.org/10.17226/25665 [4-3-2021], 238.

NUFFIELD COUNCIL ON BIOETHICS, «Genetic Editing: an ethical review», Nuffield Council on Bioethics, London 3 September 2016, in https://nuffieldbioethics.org/publications?type=report&year=2016 [19-11-2019], 128.

—, *Genome editing and human reproduction: social and ethical issues*, Nuffield Council on Bioethics, London 2018, in https://www.nuffieldbioethics.org/publications/genome-editing-and-human-reproduction [20-4-2018].

PAULLIER, J., «¿Por qué la concepción del bebé de 'tres padres' se realizó en México?», *BBC News* (2016), in https://www.bbc.com/mundo/noticias-37491942 [25-1-2020].

PEARCE, D., «The Biointelligence Explosion», in A. H. EDEN - J. H. MOOR - J. H. SØRAKER - E. STEINHART (edd.), *Singularity Hypotheses: A Scientific and Philosophical Assessment*, The

Frontiers Collection, Springer, Berlin, Heidelberg 2012, in https://doi.org/10.1007/978-3-642-32560-1_11 [21-11-2022], 199–238.

—, *The Hedonistic Imperative*, 2015.

—, «What Is It Like to Be a Philosopher?», 2022, *What Is It Like to Be a Philosopher?*, in http://www.whatisitliketobeaphilosopher.com [21-11-2022].

PEARCE, D. - VINDING, M., *Can Biotechnology Abolish Suffering?*, The Neuroethics Foundation 2017.

PERSSON, I. - SAVULESCU, J., *Unfit for the Future: The Need for Moral Enhancement*, Oxford University Press, Oxford 2012.

PEW RESEARCH CENTER'S RELIGION & PUBLIC LIFE PROJECT, «The Global Religious Landscape», 2012, *Pew Research Center*, in https://www.pewresearch.org/religion/2012/12/18/global-religious-landscape-exec/ [7-9-2022].

PICHARDO, L. M., «Reinventing Bioethics in a Post-humanist and Post-truth Society. The Present and Future of Bioethics», *Persona y Bioética* 22/2 (2018), in https://doi.org/10.5294/pebi.2018.22.2.2, 212–222.

PIEDRA, J., «Technological Enhancement and Happiness: A Review of Morphological Freedom», *Cosmos and History: The Journal of Natural and Social Philosophy* 15/1 (2019), in http://www.cosmosandhistory.org/index.php/journal/article/view/805 [13-4-2021], 275–291.

PIEPER, J., *The Four Cardinal Virtues*, University of Notre Dame Press, Notre Dame 1966.

POSTIGO, E., «Bioética y transhumanismo desde la perspectiva de la naturaleza humana», *Arbor* 195/792 (2019), a507.

—, «Naturaleza Humana y Problemas Bioéticos del Transhumanismo y el Mejoramiento Humano», in A. CORTINA - M.- À. SERRA BELTRÁN (edd.), *Humanidad infinita: Desafíos éticos de las tecnologías emergentes*, EIUNSA, Madrid 2016, 231–256.

POTO, M., «The Principle of Proportionality in Comparative Perspective», *German Law Journal* 8/9 (2007), 835–869.

POTTER, V. R., *Bioethics: bridge to the future*, Prentice-Hall, Englewood Cliffs, N.J. 1971, in http://catalog.hathitrust.org/api/volumes/oclc/226527.html [12-3-2021].

PRIETO LÓPEZ, L., *El hombre y el animal: Nuevas fronteras de la antropología*, Estudios y Ensayos. BAC. Filosofía y Ciencias, Biblioteca Autores Cristianos, Madrid 2008.

QUILL, T. E. - LO, B. - BROCK, D. W., «Palliative options of last resort: a comparison of voluntarily stopping eating and drinking, terminal sedation, physician-assisted suicide, and voluntary active euthanasia», *JAMA* 278/23 (1997), 2099–2104.

RANISCH, R., «When CRISPR Meets Fantasy: Transhumanism and the Military in the Age of Gene Editing», in W. HOFKIRCHNER - H.- J. KREOWSKI (edd.), *Transhumanism: The Proper Guide to a Posthuman Condition or a Dangerous Idea?*, Cognitive Technologies, Springer International Publishing, Cham 2021, 111–120.

REICHLIN, M., «Etica delle neuroscienze: il caso del potenziamento cognitivo.», in G. QUARANTA (ed.), *Il doping della mente. Le sfide del potenziamento cognitivo farmacologico*, EMP, Padova 2014, 19–62.

REIS- DENNIS, S., «Dignity and the Founding Myth of Bioethics», *Hastings Center Report* 53/2 (2023), in https://onlinelibrary.wiley.com/doi/abs/https://doi.org/10.1002/hast.1471 [10-5-2023], 26–35.

ROTHBLATT, M., *From Transgender to Transhuman: A Manifesto On the Freedom Of Form*, Martine Rothblatt, USA 2011.

—, *Virtually Human*, Picador Paper, New York 2015.

RUEDA, J., «Genetic enhancement, human extinction, and the best interests of posthumanity», *Bioethics* (2022), in https://onlinelibrary.wiley.com/doi/abs/https://doi.org/10.1111/bioe.13085 [13-2-2023], 1–10.

RULLI, T., «Reproductive CRISPR does not cure disease», *Bioethics* 33/9 (2019), in https://onlinelibrary.wiley.com/doi/abs/https://doi.org/10.1111/bioe.12663 [8-3-2023], 1072–1082.

SANDBERG, A., «Future of Humanity Institute 2005–2024: final report», *Future of Humanity Institute* (2024), in https://ora.ox.ac.uk/objects/uuid:8c1ab46a-061c-479d-b587-8909989e4f51 [27-5-2024], 92.

SANDBERG, A., *Morphological Freedom – Why We not just Want it, but Need it*, in (2001), in http://www.aleph.se/Nada/Texts/MorphologicalFreedom.htm [16-12-2020].

—, «Morphological Freedom – Why We Not Just Want It, but Need It», in *The Transhumanist Reader*, John Wiley & Sons, Ltd, Chichester 2013, in https://doi.org/10.1002/9781118555927.ch5 [14-4-2020], 56–64.

SANDBERG, A. - NELSON, C., «Who Should We Fear More: Biohackers, Disgruntled Postdocs, or Bad Governments? A Simple Risk Chain Model of Biorisk», *Health Security* 18/3 (2020), in https://www.liebertpub.com/doi/full/https://doi.org/10.1089/hs.2019.0115 [27-11-2020], 155–163.

SANTOS, L., «Biohacking and Transhumanism: what and why», *Relectiones* 9 (2022), in https://doi.org/10.32466/eufv-rel.2022.9.743.40-53 [11-11-2022], 40–53.

—, *Common Themes Between Humanity+ Original Documents on Transhumanism and WHO Documents on the Global Governance of Human Genome Editing. Reflexive Thematic Analysis and Implications for Bioethics.*, Ateneo Pontificio Regina Apostolorum, Rome 2023.

—, «Transhumanism and Global Governance of Human Genome Editing: Common Themes and Implications for Bioethics», *Medicina y Ética* 34/4 (2023), in https://doi.org/10.36105/mye.2023v34n4.05 [2-10-2023], 1089–1163.

SAVULESCU, J., «Procreative beneficence: why we should select the best children», *Bioethics* 15/5–6 (2001), 413–426.

SAVULESCU, J. - ALONSO, M., «Is Gene Editing Harmless? Two Arguments for Gene Editing», *The American Journal of Bioethics* 22/9 (2022), in https://doi.org/10.1080/15265161.2022.2105432 [21-11-2022], 23–28.

SAVULESCU, J. - KAHANE, G., «The Moral Obligation to Create Children with the Best Chance of the Best Life», *Bioethics* 23/5 (2009), in https://onlinelibrary.wiley.com/doi/abs/https://doi.org/10.1111/j.1467-8519.2008.00687.x [12-2-2020], 274–290.

SAVULESCU, J. - SINGER, P., «An ethical pathway for gene editing», *Bioethics* 33/2 (2019), in https://onlinelibrary.wiley.com/doi/abs/https://doi.org/10.1111/bioe.12570 [17-8-2020], 221–222.

SCHELER, M., *The Human Place in the Cosmos*, Northwestern University Press, Evanston 2009.

SEGOVIA, J. F., «La progresiva destrucción de la naturaleza y la naturaleza humana», in M. AYUSO (ed.), *¿Transhumanismo o Posthumanidad? La política y el derecho después del humanismo*, Prudentia Iuris, Marcial Pons Ediciones Jurídicas y Sociales, Madrid 2020, 49–82.

SELVARAJ, F. J., «Transhumanism and improving quality of life in the elderly: How could it shape the future of pain and suffering in the elderly?», *Medicina y Ética* 35/1 (2024), in https://doi.org/10.36105/mye.2024v35n1.02 [27-2-2024], 61–107.

SGRECCIA, E., *Manuale di bioetica. Fondamenti ed etica biomedica*, I, Vita e Pensiero, Milano 1994^2.

SNEAD, O. C., *What It Means to Be Human: The Case for the Body in Public Bioethics*, Harvard University Press, Cambridge 2020.

STOLYAROV II, G., «The United States Transhumanist Party and the Politics of Abundance», in N. LEE (ed.), *The Transhumanism Handbook*, Springer International Publishing, Cham 2019, 89–149.

THAM, J., «Resisting the Temptation of Perfection», *The National Catholic Bioethics Quarterly* 17/1 (2017), 51–62.

—, *Secularization of Bioethics-A Critical History*, Ateneo Pontificio Regina Apostolorum, Rome 2007.

—, «生命倫理學:跨文化研究: Bioethics: Cross-Cultural Explorations», *International Journal of Chinese & Comparative Philosophy of Medicine* 20/2 (2022), in https://ejournals.lib.hkbu.edu.hk/index.php/ijccpm/article/view/2238 [9-1-2023], 13–37.

THE ORGANISING COMMITTEE OF THE THIRD INTERNATIONAL SUMMIT ON HUMAN GENOME EDITING, «Statement from the Organising Committee of the Third International Summit on Human Genome Editing», 2023, in https://royalsociety.org/news/2023/03/statement-third-international-summit-human-genome-editing/ [7-3-2023].

THE WORLD MEDICAL ASSOCIATION (WMA), «Declaration of Helsinki – Ethical Principles for Medical Research Involving Human Subjects», 1964, in https://www.wma.net/policies-post/wma-declaration-of-helsinki-ethical-principles-for-medical-research-involving-human-subjects/ [28-1-2023].

TSUI, K., «Transhumanism: Meet the cyborgs and biohackers redefining beauty», *CNN* (2020), in https://www.cnn.com/style/article/david-vintiner-transhumanism/index.html [22-3-2021].

UBEL, P. A. - LOEWENSTEIN, G. - SCHWARZ, N. - SMITH, D., «Misimagining the unimaginable: the disability paradox and health care decision making», *Health Psychology: Official Journal of the Division of Health Psychology, American Psychological Association* 24/4S (2005), S57–62.

UN, «International Bill of Human Rights», *OHCHR*, in https://www.ohchr.org/en/what-are-human-rights/international-bill-human-rights [6-5-2022].

—, *The core international human rights treaties*, United Nations 2014, in https://doi.org/10.18356/9d44ec81-en [22-3-2023].

—, «Universal Declaration of Human Rights», 1948, in https://www.ohchr.org/en/human-rights/universal-declaration/translations/english [13-7-2019].

UNDP, «Global Dashboard for Vaccine Equity», 2022, *UNDP Data Futures Platform*, in https://data.undp.org/vaccine-equity/ [30-3-2022].

UNESCO, «Universal Declaration on Bioethics and Human Rights», Declaration pg.80, in http://portal.unesco.org/en/ev.php-URL_ID=31058&URL_DO=DO_TOPIC&URL_SECTION=201.html [8-6-2019].

—, «Universal Declaration on the Human Genome and Human Rights», 1997, in http://portal.unesco.org/en/ev.php-URL_ID=13177&URL_DO=DO_TOPIC&URL_SECTION=201.html [3-12-2019].

UNESCO CHAIR IN BIOETHICS AND HUMAN RIGHTS, «Bioethics, Multiculturalism and Religion», in https://www.unescobiochair.org/bioethics-multiculturalism-and-religion-3/ [15-2-2023].

U.S. TRANSHUMANIST PARTY, «Transhumanist Bill of Rights», 2015, *H+Pedia*, in https://hpluspedia.org/wiki/Transhumanist_Bill_of_Rights [9-1-2023].

—, «Transhumanist Bill of Rights – Version 3.0», 2018, in https://transhumanist-party.org/tbr-3/ [14-4-2021].

—, «U.S. Transhumanist Party – Official Website – Putting Science, Health & Technology at the Forefront of American Politics», in https://transhumanist-party.org/ [21-11-2022].

VERVE THERAPEUTICS, «Verve Therapeutics Doses First Human with an Investigational In Vivo Base Editing Medicine, VERVE-101, as a Potential Treatment for Heterozygous Familial Hypercholesterolemia», 2022, *Verve Therapeutics*, in https://ir.vervetx.com/news-releases/news-release-details/verve-therapeutics-doses-first-human-investigational-vivo-base/ [3-2-2023].

VITA- MORE, N., «About», *Natasha Vita-More PhD*, in https://natashavita-more.com/about/ [24-8-2021].

—, «H+DAO Transhumanist Decentralized Projects», 2022, *H+DAO*, in https://proposals.deepfunding.ai/graduated/accepted/f3a4c67e-06f2-4a78-90a3-853b4d69c39e [14-11-2022].

—, «History of Transhumanism», in N. LEE (ed.), *The Transhumanism Handbook*, Springer International Publishing, Cham 2019, 49–61.

—, «The Center for Transhumanist Studies», 2021, *CTS*, in https://transhumanist-studies.teachable.com/p/home [14-11-2022].

—, «The Transhumanist Manifesto (v.4)», 2020, *Humanity+*, in https://www.humanityplus.org/the-transhumanist-manifesto [11-6-2022].

—, *Transhumanism: What is it?*, Bowker 2018.

WAHBA, M. A. - BRIDWELL, L. G., «Maslow reconsidered: A review of research on the need hierarchy theory», *Organizational Behavior & Human Performance* 15 (1976), 212–240.

WEISS, T. G., *Global Governance: Why? What? Whither?*, Polity Press, Cambridge 2013.

WHO, «Ageing and health», 2022, *World Health Organization*, in https://www.who.int/news-room/fact-sheets/detail/ageing-and-health [27-2-2024].

—, «Constitution of the World Health Organization», 1946, in https://www.who.int/about/governance/constitution [29-11-2021].

—, «Human genome editing», in https://www.who.int/health-topics/human-genome-editing [12-12-2022].

—, «Statement on governance and oversight of human genome editing», 2019, in https://www.who.int/news/item/26-07-2019-statement-on-governance-and-oversight-of-human-genome-editing [13-7-2021].

WHO Expert Advisory Committee on Developing Global Standards for Governance and Oversight of Human Genome Editing, *Human genome editing: a framework for governance*, WHO, Geneva 2021, in https://www.who.int/publications/i/item/9789240030060 [13-7-2021].

—, *Human genome editing: position paper*, WHO, Geneva 2021, in https://www.who.int/publications-detail-redirect/9789240030404 [13-7-2021].

—, *Human genome editing: recommendations*, WHO, Geneva 2021, in https://www.who.int/publications-detail-redirect/9789240030381 [13-7-2021].

Young, S., *Designer Evolution: A Transhumanist Manifesto*, Prometheus, Amherst 2005.

Zettler, P. J. - Guerrini, C. J. - Sherkow, J. S., «Regulating genetic biohacking», *Science* 365/6448 (2019), in https://www.science.org/doi/https://doi.org/10.1126/science.aax3248 [14-3-2023], 34–36.

-- Ayuso, M. (ed.), *¿Transhumanismo o Posthumanidad? La política y el derecho después del humanismo*, Prudentia Iuris, Marcial Pons Ediciones Jurídicas y Sociales, Madrid 2020.

Bostrom, N. - Cirkovic, M. M. (edd.), *Global Catastrophic Risks*, Oxford 2011.

Hansell, G. R. - Grassie, W. (edd.), *H+/-: Transhumanism and Its Critics*, Metanexus Institute, Philadelphia, PA 2011.

Lee, N. (ed.), *The Transhumanism Handbook*, Springer International Publishing, Cham 2019.

More, M. - Vita- More, N. (edd.), *The Transhumanist Reader: Classical and Contemporary Essays on the Science, Technology, and Philosophy of the Human Future*, Wiley-Blackwell, Chichester 2013, in https://doi.org/10.1002/9781118555927.ch5 [14-4-2020].

Ranisch, R. - Sorgner, S. L. (edd.), *Post- and Transhumanism*, Beyond Humanism: Trans- and Posthumanism, Peter Lang, Frankfurt am Main, Berlin, Bern, Bruxelles, New York, Oxford, Wien 2015, in https://doi.org/10.3726/978-3-653-05076-9 [23-3-2021].

Savulescu, J. - Bostrom, N. (edd.), *Human Enhancement*, OUP Oxford, Oxford 2009.

ten Have, H. (ed.), *Encyclopedia of Global Bioethics*, II, Springer International Publishing, Cham 2016.

Tham, J. - Durante, C. - Gómez, A. G. (edd.), *Interreligious Perspectives on Mind, Genes and the Self: Emerging Technologies and Human Identity*, Routledge Science and Religious Series, Routledge, Oxon and New York 2019.

Weiss, T. G. - Wilkinson, R. (edd.), *International Organization and Global Governance*, Routledge, New York 2018[2].

«A Letter to Mother Nature: Amendments to the Human Constitution», 2009, in http://strategicphilosophy.blogspot.com/2009/05/its-about-ten-years-since-i-wrote.html [5-6-2020].

«Alcor - Life Extension Foundation | Cryonics», *Alcor*, in https://www.alcor.org/ [14-11-2022].

«Anders Sandberg», *University of Oxford-Faculty of Philosophy*, in https://www.philosophy.ox.ac.uk/people/anders-sandberg [14-11-2022].

Hewson, M. - Sinclair, T. J. (edd.), *Approaches to Global Governance Theory*, SUNY Press, New York 1999.

«ARRIGE – Association for Responsible Research and Innovation in Genome Editing», in https://www.arrige.org/ [7-3-2023].

«Coutrot, Jean», 2019, *Sciences Po Centre d'histoire*, in https://www.sciencespo.fr/histoire/fr/fonds-archive/coutrot-jean.html [15-11-2022].

«CRISPRcon - Conversations on Science, Society and the Future of Gene Editing», *CRISPRcon*, in https://crisprcon.org/ [5-3-2023].

«Dobbs v. Jackson Women's Health Organization 597 U.S. ___», 2022, in https://supreme.justia.com/cases/federal/us/597/19-1392/ [1-2-2023], 2228.

«Fellows», *IEET*, in https://ieet.org/fellows/ [21-11-2022].
«F. M. Esfandiary / FM-2030 Papers», New York 2007, in https://www.nypl.org/sites/default/files/archivalcollections/pdf/fm2030.pdf [14-11-2022].
«Global Citizens' Assembly on Genome Editing», *Globalca*, in https://www.globalca.org [7-3-2023].
«Home», *Oxford Uehiro Centre for Practical Ethics Home Page*, in https://www.practicalethics.ox.ac.uk/home [14-11-2022].
«Institute for Ethics and Emerging Technologies», *IEET*, in https://ieet.org/ [14-11-2022].
«James J. Hughes (University of Massachusetts, Boston)», *PhilPeople*, in https://philpeople.org/profiles/james-j-hughes [14-11-2022].
«Kurzweil», *Kurzweil. Tracking the acceleration of intelligence.*, in https://www.kurzweilai.net/ [14-11-2022].
«Martine Rothblatt», 2022, *Forbes*, in https://www.forbes.com/profile/martine-rothblatt/ [21-11-2022].
«Martine Rothblatt: My daughter, my wife, our robot, and the quest for immortality», TED Talk 2015, in https://www.ted.com/talks/martine_rothblatt_my_daughter_my_wife_our_robot_and_the_quest_for_immortality/transcript?language=nn [21-11-2022].
«Our Company | United Therapeutics», in https://www.unither.com/index [21-11-2022].
«Planned Parenthood of Southeastern Pa. v. Casey 505 U.S. 833», 1992, in https://supreme.justia.com/cases/federal/us/505/833/ [6-2-2023].
«Professor Julian Savulescu», *University of Oxford*, in https://www.practicalethics.ox.ac.uk/people/professor-julian-savulescu [14-11-2022].
«Roe v. Wade 410 U.S. 113», 1973, in https://supreme.justia.com/cases/federal/us/410/113/ [6-2-2023].
«Roe v Wade: Women travelling for abortions will be protected - Biden», *BBC News* (2022), in https://www.bbc.com/news/world-us-canada-62018206 [2-7-2022].
«Singularity University», *Singularity+iu International University of Applied Sciences*, in https://www.su.org/university?_gl=1*6m6tdf*_up*MQ..&gclid=CjwKCAiA68ebBhB-EiwALVC-NknzoOtrwaYoVKxo6iIKYjyfhLxt_3sazEtyR1hZPuA5eUSKrrRoBxoCk1wQAvD_BwE [14-11-2022].
«Terasem Movement Foundation», in https://terasemmovementfoundation.com/ [21-11-2022].
«The FreeGenes Project», *stanford.freegenes.org*, in https://stanford.freegenes.org/ [9-2-2023].
«Third International Summit on Human Genome Editing», 2023, *The Royal Society*, in https://royalsociety.org/science-events-and-lectures/2023/03/2023-human-genome-editing-summit/ [7-3-2023].
«Transhumanism's Extropy Institute - Transhumanism for a better future», *EXI*, in https://www.extropy.org/ [14-11-2022].
«Treaty of Amsterdam amending the Treaty on European Union, the Treaties establishing the European Communities and certain related acts», 1997, in http://data.europa.eu/eli/treaty/ams/sign/eng [28-1-2023], 1–144.
«Zoltan Istvan», *Zoltan Istvan*, in http://www.zoltanistvan.com/ [14-11-2022].

Index

L. Santos, *Governance of Human Gene Editing and Transhumanism*, Integrated Science 28, https://doi.org/10.1007/978-3-031-70863-3

The manufacturer's authorised representative in the EU is Springer Nature Customer Service Centre GmbH, Europaplatz 3, 69115 Heidelberg, Germany. If you have any concerns regarding our products, please contact ProductSafety@springernature.com

Printed and bound by CPI Group (UK) Ltd, Croydon, CR0 4YY
30/09/2026
02225604-0001